Writing an Effective NSF Preproposal

ISBN: 1493547062

ISBN 13: 978-1493547067

Visit **preproposalzen.com** for book orders, information about consulting, and updates on the changing landscape at NSF.

Writing an Effective NSF preproposal

Fourteen commandments to achieve preproposal zen

David W. Stephens

Dedication

I dedicate this little volume to the entire staff of NSF's division of Integrative Organismal Systems (IOS). You're awesome. More cake please.

TABLE OF CONTENTS

Preface: New strategies for a new competition

Changes at NSF

In 2012 two divisions of the National Science Foundation's Biological Science Directorate dramatically restructured the procedures for their standard grant competitions. In the Division of Environmental Biology (DEB) and the Division of Integrative Organismal Systems (IOS), investigators seeking grant funding must now submit short preliminary proposals (or preproposals for short) in January of each year. NSF organizes a review of these preliminary proposals in the spring, and in light of these reviews NSF invites investigators to submit full proposals. NSF expects to make invitation decisions in May, and full proposals are due in August. Panels review these full proposals in the fall, and program officers make funding decisions in November and December.

The big changes here are 1) preliminary proposals are now required to obtain a standard research grant in these two divisions of NSF, and 2) these divisions now have one competition per year instead of two full-proposal competitions each year. These are the big, obvious differences, but this new system changes the competition for NSF dollars in hundreds of smaller ways. The four-page preproposal is a subtly difficult document to write, and the competition for invitations is intense. How do you pack all the information you need to make your funding case into four paltry pages? The successful investigator needs new strategies and techniques that reflect the nuances of the new system.

About me

I served as program officer in the Division of Integrative Organismal Systems (IOS) in 2012, which was the first year of the preproposal system. It was a great experience in many ways, and I learned a lot about the man behind the NSF curtain. Along the way I observed firsthand many of the pitfalls of the new preproposal system. A big part of the program officer's job is to interact with the scientists who submit proposals. In doing this, I was struck with the number of misunderstandings and the amount of misinformation that exist in the scientific community that NSF serves. This isn't that surprising. NSF is as opaque as any other federal bureaucracy, but these misunderstandings hurt the scientists who seek NSF funding. This little book is, therefore, my attempt to address these misunderstandings, especially as they relate to the new preproposal process.

NSF from an applicant's perspective.

Before I arrived at NSF, I had been fairly lucky in the NSF lottery. I'd had nearly continuous funding from NSF during my twenty-five-year academic career. Of course, I'd also had my fair share of unsuccessful proposals and near misses. So while I was at NSF, I tended to see things from an applicant's perspective. My internal voice seemed to mutter, "I wish I'd known about this before I wrote my last proposal" a few times a day. NSF actually doles out a lot of advice about how to write and submit proposals, but often it emphasizes things that will streamline proposal processes at NSF. This little book tries to focus on what you — the applicant — needs to know to be successful at NSF, not just what program directors think you should know, but practical ideas about how to the turn the crank.

What's here and what's not

The NSF/BIO website is full of information about NSF's rationale for the change to the preproposal system. BIO appreciates that it's a big change, and they still want you to love them (but you don't, do you?). NSF's rationale is simply irrelevant to your strategy. The new system is here, and your academic career depends of figuring it out and making it work for you. So this guide offers practical advice about how to make it work. This book takes you linearly through the process of preparing a preproposal, from having the idea, to writing it, to responding to your reviews. It is divided into three parts: I. BEFORE YOU WRITE: What do to before you write your preproposal; II. WRITING: Strategies for writing an effective preliminary proposal; and III. AFTER THE INVITATION DECISION: How to revise your preproposal or transform it into a full proposal. Each part is further subdivided to discuss my "NSF commandments." My fourteen commandments emphasize things that applicants frequently misunderstand.

For beginners or experts?

I have tried to write so that a first-time applicant can understand my advice. To achieve this, I often give background information that will seem elementary to more senior investigators. If you're a senior investigator, you are reading this either because the new preproposal process has frustrated you, or because you're looking for a few tips that help you refine your NSF game. If you're frustrated, I encourage you to read the book from start to finish with as much of a beginner-like perspective as you can muster. You should do this because you probably need to unlearn as many things about NSF as you need to learn, because the scientific folk wisdom about NSF is full of misinformation and half-truths. Of course, many will prefer to read this book more selectively. To help with this I have organized the table of contents around my fourteen NSF commandments. Using this list, you can choose appropriate reading material to address your concerns. Please do not skip the introductory sections on choosing compelling topics of an NSF proposal. It seems very elementary—it *is* very elementary—but it is very commonly misunderstood, and it the single most important hurdle you must overcome to achieve success at NSF.

Acknowledgments

Although my experience as an NSF employee and as an applicant for NSF support has shaped the advice offered in this book, this book represents my personal opinions and nothing else.

I am grateful to the colleagues who have read earlier drafts: Craig Packer, Mark Bee, Emilie Snell-Rood, Alison Shaw and Scott Lanyon. I deeply appreciate the kindness, and collegiality that I experienced while working at NSF. NSF's employees from the program staff to the foundation leadership are hardworking and underappreciated. They don't get to make many applicants happy. They deserve much more gratitude then they receive. I am especially grateful to my two closest colleagues at NSF, Bruce Cushing and Michelle Elekonich.

Fourteen NSF Commandments

My commandments focus on high-impact actions you can take to increase the effectiveness of your preproposal specifically and more generally your portfolio of NSF applications. They repeat the major themes of the text in rule form. You should visit the indicated section of text for further explanation if "the rule" seems unclear.

Part I: Before You Write

1. One idea isn't enough; assemble a portfolio of compelling grant ideas. (Page 6)

2. Choose an irresistible enthusiasm-generating research question by asking yourself: what will completion of my proposed studies make possible? (Page 8)

> *A project that meaningfully advances science should open new topics. Making something new possible is more important at NSF than answering old questions.*

3. You need to offer a program of broader impacts that addresses NSF's societal goals with the same thoughtfulness, coherence, and focus you applied to your scientific program. (Page 12)

4. Use an award search to identify an appropriate program for your research topic and to analyze the community of panelists and reviewers for your idea. (Page 15)

Part I: Before You Write (continued)

5. Understand NSF: the rules and your multiple audiences. (Page 19)

> *The audience of program directors and program staff who will evaluate the programmatic significance of your project, and the panelists who will review and rank your preproposal.*

6. Submit two preproposals to maximize the likelihood that you will receive at least one invitation. (Page 33)

> *Make sure they are really different, or you won't be increasing your chances.*

7. Talk to your program officer on the phone. Send him or her a well-crafted summary of your idea prior to this discussion. (Page 35)

Part II: Writing

8. Write for your readers: follow a clear plan, use the active voice, and avoid acronyms. (Page 40)

9. Don't overestimate the panel's expertise or underestimate the panel's intelligence. (Page 45)

10. Use outside readers, use outside readers, use outside readers! (Page 62)

Part III: After the Invitation Decision

11. Focus on the panel summary when you read your reviews, and don't be distracted by "things that piss you off." (Page 66)

12. Take responsibility for problems in your preproposal. (Page 69)

> *Focus on what you can do to correct the misunderstanding. It serves no purpose to dwell on the supposed ignorance or malice of the reviewers.*

13. Be sure you're addressing the real problems when you revise. (Page 72)

> *Unfocused "nitpicky" criticisms often suggest that your proposal failed to generate enthusiasm in the panel. Even an ultrathorough "de-nitting" will not solve this.*

14. Learn about and exploit alternative grant mechanisms at NSF. (Page 79)

PART I:

BEFORE YOU WRITE

You need to deal with two big issues before you start writing your preproposal. First, you need at least one great fundable idea. Second, you need to understand the NSF system so that you can craft a preproposal that gives NSF what it wants. The great idea is 80 percent of the battle, so I will discuss that first, but don't underestimate the importance of understanding the nature of the competition.

Understand what makes a good idea, and have one

Commandment #1: One idea isn't enough; develop a portfolio compelling of grant ideas.

Your scientific training tells you what a good idea in your area is, right? Obviously this is partly true. Your preparation and your resources define the sort of questions your research addresses. To be successful at NSF, you need to excite your peers about your research, and the peers in question are reviewers, panelists, and program directors. So when you sort through your list of good ideas (and I hope you have a long list), you should evaluate them against this standard.

Obviously, I cannot tell you whether a project on honeybee genomics is better than a project on juvenile hormone. Instead, I offer a series of thoughts about what makes a good idea for a preproposal and what does not.

Developing your portfolio of grant ideas

One of the revelations I experienced during my time as a program officer is that those who are successful at NSF—and I'm talking about the top 1 to 2 percent of NSF grant-getters—are incredibly flexible. They have lots of project ideas, and they have a knack for crafting their ideas to fit the ever-changing landscape of NSF programs and priorities. In contrast, I have always had one or maybe two potentially fundable ideas on the boil at any one time. Yet if we measure success simply by fundability (and we probably shouldn't!), this is really a no-brainer strategy. The successful principal investigator (PI) submits lots of proposals to many different programs and crafts these proposals in direct response to what NSF wants.

Although the pursuit-of-knowledge purist in me objects to this on philosophical grounds, if your promotion, your raises, and your quality of life depends on obtaining grants, you can at least move in this direction.

Brainstorm your idea portfolio

I suggest that you start by making a list of your best grant ideas. I like the brainstorming and mind-mapping techniques for assembling this list. The key ideas here are 1) you should not edit yourself — put down everything that comes to you; and 2) you should use lines, arrows, and annotations liberally to show the connections between your ideas. This is a completely free-form exercise, so how you do it is totally up to you. But do try to get as many ideas as possible down, and be sure that at least some of these ideas are on the crazy edge of your field. In the right situation, NSF loves risky projects. Take this brainstorming process seriously, and give it some time to percolate (don't dash it off in twenty minutes).

Once you've put down your list of all possible ideas, it's time to pare it down. You might completely cross off ideas that are impractical, but don't be too aggressive about deleting things. As you prune, remember that the advantage of a portfolio over a single idea is that it spreads risk, so a portfolio of ten very similar ideas defeats the purpose of this exercise. So as you review your list, be sure that the items vary along several dimensions; they might differ in the hypotheses tested, the empirical approach, the level of biological organization addressed, and so on.

Poll your colleagues

A successful proposal needs to generate enthusiasm at NSF, especially among panelists and program officers. Generating enthusiasm begins with your idea. Some ideas just seem sexier than others. I suggest that you review your "idea list" with colleagues and graduate students to get some feedback on which ideas are natural enthusiasm generators.

In doing this, you should try to get a diversity of opinions. You want an evaluation from people who best represent your target program. Typically this means that they are in your broad area but not in your subspecialty. So look for people at the edges of your field.

Commandment #2: Choose an irresistible enthusiasm-generating research question by asking yourself: what will completion of my proposed studies make possible?

Things that seem like good ideas but aren't

Now that you have your list of ideas, it's time to focus on one. In the next few paragraphs, I give some general advice about good and bad ideas.

As some readers will know, the explanation of elaborate sexual ornaments in male animals is a long-standing question that could be explained by any several difficult-to-separate hypotheses. You would naturally think that this textbook, unsolved problem in behavior and evolution would be great topic for a project. You'd be wrong. You'd be wrong because in the world of NSF a problem that people have been working on for years without resolution seems like a dead end.

The point here is not to denigrate studies of sexual selection, but to make the important point that long-standing unsolved problems are a mixed blessing in an NSF question. They can be very powerful if you offer a way to get out of the rut, and especially if your rut-escaping idea will lead to something totally new, but simply studying this problem in another species or another environment is not a promising idea.

A proviso here is that these sorts of longtime-problem projects do tend to play better with panelists than with program directors. Program directors need to justify their decisions in light of their current portfolio of projects, and if they are already funding projects on your long-standing problem, that's a big negative for your preproposal.

Three fatal flaws: incremental, me-too, stamp-collecting

In halls of the NSF, people discuss proposals: what's good, what's bad, what's exciting, and what's not so exciting. To help you understand what makes a good project idea, here are three things that you don't want people to say about your project:

"It's incremental" means that NSF sees your idea as a relatively minor linear extension of what is already being done in your field, or even in your own lab.

'It's me-too science" means that NSF sees the proposed project as largely imitative of some other research program that is already established.

"It's stamp-collecting" means that NSF sees your project as doing almost exactly what someone else has done in another species or system.

You get the idea. Remember that these are subjective and relative judgments. One program officer's "fatally incremental" could be another's "well, it's a bit incremental, but it still has merit because…" What you need to do with this information is craft your proposal and ideas to avoid these pitfalls. Every proposal needs to make a strong case that it is *not* incremental.

Have a good answer to the "what becomes possible" question

A competitive preproposal (or proposal) needs to make a strong argument that **something new and significant will**

become possible when the proposed research is completed. This ought to be "a duh," but clearly it isn't to many applicants. It ought be "a duh" because when we speak of "advancing science," we surely mean that we answer questions and that these answers *make new questions* possible. Many applicants seem to stop at the "answering questions" and miss the key point that the importance of these answers depends on the new things that emerge from the answers.

OK, stop and reread the preceding paragraph. If you reconsider it carefully, you may find that it contradicts many of your preconceptions about what makes science important. Developing an example that will be cited in future textbooks, testing a long-standing hypothesis, publishing results in *Science* and *Nature*—these are all nice things, and they impress people at NSF. They are, however, trivial details when compared to opening new fields and making new scientific questions possible. This is what your work should to be trying to do, and the potential contribution of your research to making new science possible is the strongest argument you can make at NSF. Why? NSF wants to shape and catalyze American science, and a project that changes the kind of studies that scientists propose is a billion dollars' worth of effect for a half-million-dollar investment—exactly what NSF wants.

Applying this standard, we might ask ourselves what kinds of projects will—in the long run—change the way science is done. There are two broad categories. First, studies that develop new methods have a pretty straightforward claim about making new things possible. Some investigators who think of themselves as "pure" scientists might look down their noses at the development of new methods, but everyone working in science must sometimes develop or refine methods, so you should not overlook "what becomes possible arguments" that flow from methodological innovation in your work.

The second broad category of research that makes new things possible is "paradigm-changing" research. If your research overthrows a traditional approach and offers a novel way to ask questions and generate hypotheses, then you may have an argument for making new things possible. This category is a little more difficult to convincingly sell than the methodological category, because recognizing paradigm-shifting research before the paradigm has actually shifted requires a subjective judgment. In addition, you have to careful about overstatements when you're making a claim like this. Don't claim to offer a new paradigm for biology when you're working on synaptic transmission. Many well-crafted NSF projects will have some "paradigm-shifting" potential and some innovative methods. You can use these basic elements to craft your own compelling argument about what your research makes possible.

This is the magic bullet. If you have good answer to the "what becomes possible" question, you have the keys to the kingdom, which is to say you have the keys to generating enthusiasm for your proposal among both panelists and program officers.

How important is preliminary data?

Historically, investigators have been able to greatly strengthen their grant proposals by adding preliminary results. A graph or a table of preliminary results can wordlessly make the case for the importance and feasibility of your project. As you decide which of the projects from your portfolio to develop as a preproposal, you should obviously favor those where you have some preliminary data. There are, however, a couple of issues to consider here. First, NSF has said that the preproposal system will de-emphasize preliminary data. According to this view, preproposals should emphasize conceptual issues and leave the details for the yet-to-be-invited full proposal. My experience with preproposal panels suggests that you ignore the issues of feasibility at your peril. Second, the severity of this issue depends on exactly what you propose. If you will be using techniques that are well known and well used in your lab, then you might be able to de-emphasize preliminary data as NSF suggests, but if your idea is surprising or little known, you had better include at least a little preliminary data, or the panel will miss it.

Commandment #3: You need to offer a program of broader impacts that addresses NSF's societal goals with the same thoughtfulness, coherence, and focus you applied to your scientific program.

So far, I have focused on how to choose a strong scientific question. In NSF-speak, I have emphasized the "intellectual merit" review criterion (which is, of course, central to NSF's mission and your success or failure at NSF). As most readers will know, NSF's second review criterion is "broader impacts," and you need to spend some time and energy developing clear ideas about the broader impacts of your proposed project. In practice, you can only devote a small part of your preproposal to broader impacts (typically a half page to one page), but you still must articulate a thoughtful plan to achieve "broader impacts."

I offer several key points gleaned from the reactions of panelists and program directors to a wide range of proposed broader impacts.

You must develop a broader-impacts program that goes beyond the things you (and other researchers) normally do. For example, a broader-impacts section explaining that you will "train graduate students and publish your results" is a nonstarter. This is not because training graduate students is not a legitimate broader impact (it is), but because everyone does this, and it will not be competitive. If space allows, and you have an especially creative or remarkable program for graduate training, you should feel free to mention it.

Make your proposal forward looking, not backward looking. Even if you have an amazing track record of student training, public outreach, and engaging students from underrepresented groups in science, you must outline a program for the broader impacts that WILL result from your project. You can mention your track record in passing, but do not depend on it.

A competitive program of broader impacts should be just that: a well-integrated *program* of activities that promote NSF's societal goals. The research you proposed represents a thoughtful and coherent attack on a significant scientific question. Your program of broader impacts needs the same thoughtfulness, coherence, and focus.

Although it is not technically required, reviewers prefer programs of broader impacts that are well integrated with the proposed research.

If you propose a training component, don't forget to also include a mechanism to assess the success of your plan.

Now that you have these principles in mind, consider the National Science Board's (NSF's ruling body) recent statement about broader impacts. The board urges reviewers to ask:

How well does the activity advance discovery and understanding while promoting teaching, training, and learning? How well does the proposed activity broaden the participation of underrepresented groups (e.g., gender, ethnicity, geographic, etc.)? To what extent will it enhance the infrastructure for research and education, such as facilities, instrumentation, networks, and partnerships? Will the results be disseminated broadly to enhance scientific and technological understanding? What may be the benefits of the proposed activity to society?

I recommend that you brainstorm about the broader impacts of your project with this list in mind:

1) Promoting teaching, training, and learning. As mentioned above, a training program that is limited to the conventional activities of research labs (graduate students and postdocs) will not be competitive. If you propose a training program as part of your broader impacts, you need to focus on something creative and out of the ordinary. A common path is to involve undergraduates or high school students from underrepresented groups.

2) Broaden the participation of underrepresented groups in science. Panels typically look very favorably on credible efforts to involve students from underrepresented groups. The most common pitfall is a lack of credibility. You will need to be specific about how you will recruit these participants.

3) Enhance infrastructure for research and education. I have seldom seen this argument made successfully. As with graduate student training, building the infrastructure for our research is a normal and expected part of what we do. Again, as with training, you would need to offer a program of infrastructural development that is unusually creative and beneficial to other investigators and educators.

4) Broadly disseminate your results to increase scientific understanding. This does NOT mean publish papers. It's interpreted to mean disseminate results to the public or to intellectually distinct disciplines. Although many investigators have very strong track records of broad dissemination (e.g., writing popular books, appearing on TV shows), it can be difficult to translate these track records into a tightly coordinated program of research and public dissemination.

Do not make a broader impacts checklist. You not should contrive a way to cover all these points in your broader impacts. In fact, the National Science Board explicitly discourages this checklist approach to broader impacts. Instead, I recommend that you develop a program that draws together two or three of these broader-impact components. You might, for example, propose a program to recruit students from underrepresented groups to your project (goal two) and have them work with a documentary filmmaker to produce a film about some accessible aspect of your work (goal four). In my experience, many strong broader-impacts programs capitalize on tools that are already available to the investigator. Proposals from urban areas might develop a program emphasizing broadening participation by working with inner-city high schools, but it's just as plausible that a proposal from a less populated area could make an impressive case for bringing science to rural school kids.

> *Commandment #4: Use an award search to identify an appropriate program for your research topic and to analyze the community of panelists and reviewers for your idea.*

Understand what your program wants

When sorting through your list of preproposal topics you should keep a weather eye on the fit between the NSF program you're targeting and your project idea. There are basically two ways to get authoritative information about the sort of thing your program is eager to support: an award search and talking to your program officer.

Exploit the information available in an award search

I recommend doing an award search first. This gives you a snapshot of the target program's current portfolio. You do this from the NSF website. To understand award searches, you need to know a little about NSF's structure, because this information will help you find what you're looking for. The BIO directorate is organized into divisions, and two these — IOS (Integrative Organismal Systems) and DEB (Division of Environmental Biology) — have instituted the preproposal system we're focusing on. Each of these divisions is subdivided further into subject area **clusters**. For example, IOS has clusters for Neural Systems, Behavioral Systems, Physiological and Structural Systems, and Developmental Systems. Clusters are further subdivided into **programs**. For example, the Neural Systems cluster runs programs in "Organization," "Activation," and "Modulation."

Programs are important to you because they represent the pots of money that your preproposal is competing for, and typically each program organizes its own annual competition and runs it own review panel or panels. It follows that if we want to use award search to identify the subject matter for a given competition and to characterize the potential panelists, we want to see what a program has funded.

You can use this information to do an award search in two ways. First, each cluster has a landing page on the NSF website. If you're an IOS investigator, you can go the main IOS page at **http://www.nsf.gov/div/index.jsp?div=IOS** and click on one the four links for a cluster description about halfway down the page. There is useful information on this page about who your program officer is and how to contact him or her, but for our purposes the important bit is at the bottom. Each cluster description has a link that offers a cluster-level award search. Click it!

What Has Been Funded (Recent Awards Made Through This Program, with Abstracts)

This will generate a list of funded projects that looks like this:

IOS: The Adaptive Significance of Juvenile Coloration: Precocial Partner Preference
Award Number:1257881; Principal Investigator:Stanley Fox; Co-Principal Investigator:Ronald Van Den Bussche, Matthew Lovern, Jennifer Grindstaff; Organization:Oklahoma State University;NSF Organization:IOS Award Date:09/15/2013; Award Amount:$500,000.00; Relevance:96.0;

Adaptive distribution of morphological specialists in social insects: New insights into the evolution of division of labor
Award Number:0841756; Principal Investigator:Anna Dornhaus; Co-Principal Investigator:Scott Powell; Organization:University of Arizona;NSF

Organization:IOS Award Date:06/15/2009; Award Amount:$450,000.00; Relevance:96.0;

Dissertation Research: Biomechanics of Feeding in Loggerhead Shrikes Award Number:1110716; Principal Investigator:Margaret Rubega; Co-Principal Investigator:Diego Sustaita; Organization:University of Connecticut;NSF Organization:IOS Award Date:06/01/2011; Award Amount:$14,452.00; Relevance:96.0;

Collaborative Research: Plumage redness and good genes in the House Finch Award Number:0923600; Principal Investigator:Geoffrey Hill; Co-Principal Investigator:; Organization:Auburn University;NSF Organization:IOS Award Date:08/15/2009; Award Amount:$300,000.00; Relevance:96.0;

Dissertation Research: Multiple Paternity in P. antipodarum, a New Zealand Snail   Species
Award Number:1110396; Principal Investigator:Curtis Lively; Co-Principal Investigator:Deanna Soper; Organization:Indiana University;NSF Organization:IOS Award Date:06/15/2011; Award Amount:$12,266.00; Relevance:96.0;

DISSERTATION RESEARCH: Immune-based Maternal Effects: Investigating Variation and Plasticity in Female Deposition of Protective Immune Compounds   into   Eggs
Award Number:1110563; Principal Investigator:David Winkler; Co-Principal Investigator:Esther Angert, Anna Forsman; Organization:Cornell University;NSF Organization:IOS Award Date:06/01/2011; Award Amount:$14,953.00; Relevance:96.0;

I will have a bit more to say about how to use this information later on. Now, however, we need to discuss how you can get a more refined and focused list. The problem with this is that you will often want to delve below the cluster level to the program because that will be the pool that identifies the appropriate topics' potential reviewers. Table 1 gives a list of program element codes for DEB and IOS. You can use these to get more precise information about the competition that's most immediately relevant to you. To do this, go to the main NSF page, www.nsf.gov, and choose "AWARDS" from the tabs across the top. This opens a menu, and you should choose the "Search Awards" option from this menu. Now choose the "advanced search" option from the top banner, and this will reveal a more complicated set of options. Lo and behold, you will see a box labeled "element code." Enter the appropriate element code, and you will see what your program has funded recently.

Table 1. Program Element Codes

Division	Cluster	Program/Panel	Element Code
IOS	Physiological and Structural Systems (PSS)	Symbiosis, Defense and Self-recognition (SDS)	7656
IOS	Physiological and Structural Systems (PSS)	Physiological Mechanisms and Biomechanics (PMB)	7658
IOS	Physiological and Strutural Systems (PSS)	Integrative Ecological Physiology Program (IEP)	7657
IOS	Neural Systems	Organization	7712
IOS	Neural Systems	Activitation	7713
IOS	Neural Systems	Modulation	7714
IOS	Behavioral Systems	Animal Behavior	7659
IOS	Developmental Systems	Plant, Fungal and Microbial Developmental Mechanisms Program	1118
IOS	Developmental Systems	Animal Developmental Mechanisms Program	1119
IOS	Developmental Systems	Evolution of Developmental Mechanism Program	1080
DEB	Ecosystem Science	Ecosystem Studies	1181
DEB	Ecosystem Science	Long term ecological research	1195
DEB	Ecosystem Science	Ecosystem Science Cluster	7381
DEB	Evolutionary Processes (EP)	Evolutionary Processes Cluster	1127
DEB	Evolutionary Processes (EP)	Evolutionary Ecology	7377
DEB	Evolutionary Processes (EP)	Evolutionary Genetics	7378
DEB	Population and Community Ecology (PCE)	Population and Community Ecology Cluster	1128
DEB	Population and Community Ecology (PCE)	Population and Community Ecology Program	1182
DEB	Population and Community Ecology (PCE)	Long-term Research in Environmental Biology	1192
DEB	Systematics and Biodiversity Science (SBS)	Systematics and Biodiversity Science (Cluster wide)	7374
DEB	Systematics and Biodiversity Science (SBS)	Phylogentic Systematics	1171
DEB	Systematics and Biodiversity Science (SBS)	Biodiversity: discovery and analysis	1198
DEB	Systematics and Biodiversity Science (SBS)	Advancing Revisionary Taxonomy and Systematics	7375
DEB	Systematics and Biodiversity Science (SBS)	Partnerships for Enhancing Expertise in Taxonomy (PEET)	7376

What you'll find and how to interpret it.

Once you see the list of things your program is currently funding, you can peruse the names of investigators, the titles of projects, and the duration and amount of the award. In addition, you can click a link to see the public abstract (more about that later). The award search for your target program is a gold mine of information. (In fact, some energetic applicants download this information to Excel for sorting, filtering, and pie charting!) It gives you the "who," "what," and "how much" details that are critical to your success at NSF:

1) The investigators (PIs in NSF lingo) represent potential panelists, and moreover this list tells you who NSF thinks the "community" for this program is.

2) The project titles and abstracts give you the clearest possible picture of the topics the program covers. You should think of this as a way to discover the intellectual center of mass of the program, but it *does not* follow that you should craft a proposal that mimics an existing award. Specifically, you should abandon any plans you have to propose a "me-too" project that closely resembles any single project in your program's award list.

3) The award list gives you an idea of the normal size and duration of awards in your program, and this will useful in planning your project.

In most cases, you will know which program is appropriate for your work, and you can clearly target this program as I describe here. Of course, some investigators have project ideas that may be appropriate for many different programs in NSF's BIO directorate. In this case, you obviously can (and should!) use an award search to find the right home for your project. I will have more to say about applying to multiple programs elsewhere.

Commandment #5: Understand NSF: the rules

and your multiple audiences.

Understanding NSF's system

Once you have settled on the basic idea you will propose, you need to make sure that you understand the NSF system before you begin to write. This is important because it affects what you should emphasize about your idea. NSF is a large bureaucracy with many moving parts, so there's an enormous amount to know. In this section, I focus on key points that are frequently misunderstood and that are critical to your success.

Make peace with the preproposal system

The scientific communities served by the Division of Environmental Biology and Integrative Organismal Systems have experienced considerable "discomfort" and even anger about the new one-competition-per-year, preproposal-required system. Some of my colleagues feel very passionately that this change does not serve the interests of science. As a point of American science policy, this is an interesting and even important debate. However, as individual who seeks NSF support, it's irrelevant.

The preproposal system is here now, and your career in science depends on making the system we now have work. It's important that you turn your energies toward understanding this new system. My experience as a program officer suggests that applicants who take a "complain and blame" approach to NSF are much less successful than applicants who approach programs with a "help me figure this out" attitude. So I advise you to act like a successful applicant and focus on understanding the NSF system and how to make it work for you.

Know the rules

NSF is a government agency, and it takes its rules and regulations very seriously. NSF is a much more regulated environment than the universities where most of us work. What this means for you as an applicant is that you must understand the rules that apply to your application. At the lowest level, NSF-funding mechanisms are governed by a *solicitation*, so you need to familiarize yourself with solicitation that's relevant to your application. At a higher level, NSF's Grant Proposal Guide (the infamous GPG) gives the generic specifications for NSF proposals. If the relevant solicitation is silent on a given topic (the allowable typeface, for example), then you should follow the GPG's rules.

To prepare for writing a preproposal, I recommend that you familiarize yourself with three key documents. You need to read them before you write, but you will also need to refer to them as you work on your preproposal. I usually make an "NSF pubs" folder on my desktop and download the current pdfs of these three documents into it. Here's my required reading list:

1) **The Grant Proposal Guide.** Be sure to download the current Grant Proposal Guide. For example, if you Google something like "NSF GPG," you will find a bazillion copies of out-of-date NSF documents, so go straight to the source at www.nsf.gov. At this writing, the current GPG is 13-1 and it's available at http://www.nsf.gov/publications/pub_summ.jsp?ods_key=nsf13001.

Note that the GPG is now part of a larger NSF document called the PAPPG (pronounced *pap-g*), which combines grant preparation information (the GPG) with the rules for administering award (the old PAM, proposal and award manual). For the purposes of proposal preparation, you need only the GPG, and you can download this separately.

2) **Your target program's solicitation**. In the case of the preproposal system, the rules depend on whether you plan to

apply to the Division of Environmental Biology (DEB) or the Division of Integrative Organismal Systems (IOS). There is some effort to keep the two solicitations in sync, but there are real differences, so you need to read the one that's relevant to your application.

If your target field is Organismal Biology (animal behavior, neuroscience, physiology, or development), you should download Solicitation 13-600 from

http://www.nsf.gov/publications/pub_summ.jsp?WT.z_pim s_id=503623&ods_key=nsf13600

into your "NSF pubs" folder.

If your target field is Environmental Biology (evolution, ecology, or systematics), you should download solicitation 14-503 from

http://www.nsf.gov/publications/pub_summ.jsp?WT.z_pim s_id=503634&ods_key=nsf14503.

Notice that both solicitations are new for the 2014 competition. My website pre-proposalzen.com explains the changes, but they do not affect pre-proposal preparation.

3) **NSF's revised merit review criteria.** In late 2011, the National Science Board (NSF's governing body) revised and amplified the long-standing merit review criteria of "intellectual merit" and "broader impacts." These revised criteria are fully reflected in the latest revisions of the GPG/PAPPG (13-1). Get this document from

http://www.nsf.gov/nsb/publications/2011/meritreviewcriter ia.pdf.

The new criteria are an improvement in the sense that they give specific guidance on the interpretation of these two rather vague terms. Moreover, this new document breaks the two existing criteria into components.

These revised criteria are very new, but you should keep them in mind for the following reason. National Science Board criteria have a way of ending up on the evaluation forms that reviewers, panelists, and program officers must complete when they justify their favorable opinion about your proposal. You can make their job easier by taking every opportunity to address these criteria in your written proposals.

A comment on reading NSF documents

Once you have downloaded these three documents, you need to set aside some time — ideally well before you start to write — to read and review them. For me, at least, this best done on a quiet weekend with a highlighter and a coffee. You need the highlighter because some of the information is these documents is absolutely critical to getting the details of your proposal right, but a great deal of it is government boilerplate. You should be able to learn pretty quickly what can be skimmed and what needs your careful attention.

Obviously, as you read the GPG and solicitation you should be careful to note any places where they differ. Remember that the relevant solicitation overrules the GPG anywhere that they differ, but the GPG governs on any matter where the solicitation is silent. So pay attention, and keep that highlighter (whether it is digital or analog) busy.

Know the procedures

The solicitation and GPG provide the formal rules that your submission must follow, but it's very useful to understand what will happen to your preproposal after it arrives at NSF. This is important because the text you write can influence what happens to your preproposal at a few critical choice points.

When your institutional representative submits your preproposal, it starts a series of events that will determine the success or failure of your project. Two things happen right away. First, FastLane generates a permanent ID number for your project. These are seven-digit numbers that begin with two digits representing the federal fiscal year in which your proposal was submitted. This number is important to you because it is NSF's ultimate identifier of your proposed project. You should include it in the subject line of any e-mail you send to NSF about this project. Second, your institutional representative's submission moves your preproposal into the "compliance checking" phase of the NSF review process.

Compliance checking

Immediately after the solicitation deadline, the NSF staff runs a compliance review for all the submissions in each division. Some of this is automated, and some is done manually. The compliance checking process will check your preproposal for all the required elements (as specified in the solicitation's preproposal preparation checklist), and it will also check for extra, unwanted material. For example, if you incorrectly submitted a budget, the system will catch this.

If you're preproposal is not compliant, NSF *can* return it without review ("'RWR it" in NSF terminology). If this happens, it is the end of the line for your idea, at least for this year. This is one reason why it's so important to know and follow the rules. In most cases, however, NSF will send you an e-mail and ask you to fix the issue. You typically have five days to submit a revision via FastLane's file update mechanism. If you miss this, again it's the end of the line for your proposal. Occasionally, applicants leave on vacation or research trips immediately after they submit things to NSF. It's very important that you stay in at least e-mail contact for the first week or two after you submit anything important to NSF.

If your preproposal is compliant (and if you pay even minimal attention to the rules you'll probably be OK), it is assigned a program officer within the program you designated on the cover page. Your preproposal will be transferred to the internal version of FastLane, which is called eJacket. In eJacket your proposal has an electronic jacket or file that has links to all the relevant information about your proposal, including what you've submitted, internal correspondence about your proposal, your reviews, the identities of the reviewers, and so on. You've probably guessed that your proposal ID number is the number that identifies your electronic jacket.

So now the program director is looking at one hundred to three hundred preproposals in his or her "My Work" page, and her first thought is, "Do these really belong in my program?" The program director will now work to reduce this list in two ways. First, she will look for proposals that are simply inappropriate for the division or NSF in general. For example, a purely clinical proposal is inappropriate for NSF because that's NIH's job, or a proposal that focuses on a purely computational problem could be inappropriate for the entire BIO directorate. These inappropriate proposals will be returned without review.

For most of us this is pretty unlikely. A much bigger is issue is "horse trading." As the program officer reviews her "My Work" pile, she sees a few proposals whose titles suggest that they would fit better in another program. She will open the jacket and check the project summary. Suppose, for example, that a preproposal on keystone predators has been submitted to the neuroscience program in IOS. The neuroscience program feels that it's a better fit to community ecology, so they send an e-mail with the proposal ID number to the program director in community ecology: "Please look at 1345679. It was submitted to the neuro cluster, but I think it's a better fit to your program." If the community ecology program director agrees, the proposal is moved from neuroscience and will be reviewed in community ecology. Programs vary in how much they tell the applicant about this. DEB tends to ask the applicant for permission to do this (although the request has an "allow the transfer or die" tone to it), whereas IOS does not.

The effect of this horse trading can be positive for some applicants, but usually it's a bad thing from the applicant's perspective. For the clueless applicant, this can be OK. If the clueless applicant submitted his proposal to the wrong program in the first place or didn't take the time to do the groundwork about "program fit," which I've suggested previously, then he might get a more appropriate review. However, for the applicant who has thoughtfully crafted his proposal for "the audience" represented by a particular program, this is an unmitigated disaster. Nothing is more important than writing for the correct audience, and figuring out the NSF audience is hard enough without being horse-traded to another program. Finally, this can cause compliance problems because the solicitations limit the number of preproposals you can submit to a division (two to IOS and two to DEB). If the hapless applicant in our example above had already submitted two preproposals to DEB, he would be over the limit when community ecology accepted the transfer from neuroscience (which is in IOS). It's not 100 percent certain what would happen in this case, but in theory DEB could return one the DEB preproposals without review.

I hope it's clear that you really need to do your homework about program fit before this happens to you.

Panel construction

Once the program director or directors have settled on the list of preproposals that will go forward into the current round of competition, they will begin to assemble a panel. Your proposal has relatively little to do with this process except that the program directors will try to ensure that the expertise needed to evaluate your idea is represented on the panel. However, this is only one of the many variables the program director needs to consider. NSF's high-level administration must approve each panel, and they will look for geographical distribution of panelists from across the United States, a mix of junior and senior scientists, scientists from primarily undergraduate institutions as well as from "Research 1" universities, and so on. After the expertise and diversity issues are attended to, the program director is empowered, even encouraged, to choose panelists who she feels represent important new directions in the field. This is relatively small influence given all the constraints, but it's one of tools that the program director has to "lead" his or her field.

Review procedures, panel only, no co-review

The review procedures for preproposals differ from the way NSF reviewed full proposals under the old system in two very significant ways. First, there are no external reviews of preproposals; instead, all the reviews you see will come from panelists. You will normally see three reviews, one from each of the panelists assigned to your proposal. Second, NSF does not do co-reviews for preproposals. In a co-review, two or more programs consider your project, doing reviews, sitting in on each other's panel discussions, and so on. NSF sees co-review as way to promote interdisciplinary research, because it allows them to get expert opinion from different perspective.

These two facts should influence your preproposal writing strategy. First, the absence of external reviews means that you must write your preproposal focusing on the ideas and opinions of your field's "everyman." You cannot depend on the praise you might receive from a well-respected expert in your field. You should think instead of the kind of colleague who teaches the introductory course in your field. Ask yourself what is the common base of knowledge these types of scientists will have.

The absence of co-reviews can change your submission strategy in several ways. To see this, consider the way co-reviews typically play out. Commonly, the ideas that are just outside of panel A's expertise seem novel and generate considerable enthusiasm, but ideas that seem novel to panel A are firmly in panel B's comfort zone, and they are often found wanting on some technical grounds. Panel B's more technical critique trumps panel A's enthusiasm, and the proposal goes down in flames. I think this means that the questions addressed by your preproposal need to be pretty close the "center of mass" or your target panel's expertise, but you have a little more leeway to propose techniques and approaches that are outside the mainstream for your program. This will often generate enthusiasm. (Of course, you may have to pay the piper when it comes time for a full proposal where co-reviews are very much in play.)

While co-reviews are irrelevant or even negative for most investigators, a few investigators in the IOS and DEB community depend on the co-review system because of the interdisciplinary nature of their research. If you fall in this category, you need to focus your preproposal so that it clearly addresses a core issue (or issues) in your target program. You will be able to expand on the interdisciplinary nature of your research when your full proposal is invited, but it's critical at the preproposal stage that you focus on a topic that "makes some new and exciting thing possible" within the subject area of your target program.

This section gives a brief review of what happens during an NSF panel. You may or may not need to know these details, but I think it will help to understand the procedural gauntlet that your preproposal must run. I offer this overview with some reservations, because NSF panels vary a lot from one program to the next. So these details represent basic procedures that may vary from one panel to the next.

Prepanel procedures

Depending on the proposal load, each NSF program will assemble one or more panels of experts in your field. A program with two hundred or so preproposals might organize as many as three panels. Although the numbers vary, panels range from eight to twenty panelists, and each panelist is usually assigned to evaluate about twenty proposals. And typically, three panelists will read each preproposal. The three panelists assigned to your preproposal will have different roles: primary, secondary, and tertiary. Each panelist's reviewing load will be equally divided among the primary, secondary, and tertiary roles. Before NSF assigns preproposals to panelists, it typically sends them a survey and asks them to indicate which preproposals they most want to review. The program uses this information to make assignments and then to generate a "batting order" or agenda that governs the order in which the panel will discuss your proposal.

After the program makes the panelists' assignments, the panelists have about a month to prepare and submit written reviews. These reviews must specifically address the intellectual merit and broader-impact review criteria. In addition, the panelists must provide a rating on the following scale; E = excellent, V = very good, G = good, F = fair, and P = poor. Panelists can give a split rating such as E/V to indicate, for example, a rating between excellent and very good. The panelist submits his or her reviews via the FastLane panel system, and once the panelist has submitted all of her assigned reviews, she can see the reviews submitted by other panelists.

Panel meeting procedures

When your preproposal comes up in the batting order, the program director (who chairs the panel) will announce your preproposal, and the primary panelist will begin the discussion. The primary panelist will give a brief overview of the proposal and then offer his or her assessment of the intellectual merits and broader impacts of the proposed project. The agenda usually allots about five minutes for the primary panelist's remarks. The secondary panelist is then asked to offer his or her comments and is often instructed to emphasize points of disagreement with the primary panelist, usually about three to four minutes. After this, the program director asks the tertiary panelist to contribute, and finally the program director will ask the panel at large for discussion. Unassigned panelists may ask for clarification or even defend or attack a claim made by one of the assigned panelists. The full text of your preproposal is available electronically to the unassigned panelists, and in theory they can read it ahead of time, but more often they skim it during the discussion.

Although the actual assignment varies, one of the assigned panelists will be the scribe. The scribe is formally responsible for drafting the panel summary. Often this is the tertiary panelist, but sometimes it is the primary. Similarly, one of the panelists, often the secondary, may be assigned the job of stating the ratings and summarizing the salient points of the reviews.

The program director is strictly neutral during the discussion of your preproposal but may ask for further clarification. It's especially common for the director to remind panelists to discuss broader impacts, for example. In rare occasions, panelists may do something "naughty" such as allowing something that's not strictly in the proposal to enter into the discussion, and the program director should correct this and advise the panel to avoid these transgressions.

The ranking board

At the end of the discussion, the program director asks the panel to place your preproposal on the ranking board. A screen at the front of the panel room displays the ranking board constantly, and it is the focal point of the discussion. The boards vary a bit. Two are in wide use. One uses Excel to put the proposal in four columns, and the other uses PowerPoint to move metaphorical Post-it notes around the board in a free-form way. In both cases there are four categories, crudely corresponding to high quality, medium quality, low quality, and not competitive (NSF argues endlessly about the names for these categories, but it really doesn't matter to you). Some programs rank the proposals within each category so that they have a record showing that "Smith in High" was actually higher than "Jones in High," for instance, but other programs do not. Program directors often struggle with panels to keep a fairly even distribution across the board. Panels often put too many in high quality, but occasionally they will put too many in not competitive. These uneven distributions are problematic because they reduce the value of the panel's advice. If everything is in one category, the program director will have the final say in what gets invited and what's declined. You will eventually find out which category the panel placed your preproposal in, but you will seldom discover anything about your ranking within a category. Program directors typically guard this information jealously.

The program director will ask the panel to review the board at several points during the panel meeting, and at this time the panelist may move proposals up or down on the board. This is obviously important because relative rankings solidify as the panel begins to see the whole distribution of proposals.

Remains of the day

Your preproposal will have its moment in the sun for somewhere between eight to fifteen minutes. This discussion produces two things: the written panel summary and your preproposal's ranking on the board. In addition, you will eventually see the separate written reviews of the three panelists and, of course, their ratings. Later I will discuss how to decode this information. For now, however, the thing you need to know is that the three panelists assigned to your preproposal are your target audience. It's virtually impossible to know exactly who these panelists will be, but your preproposal must generate enthusiasm from this select audience if it is going to be successful. The key to this is to write for the well-informed but "generic" member of your target program's community, and to make a clear argument about the new and significant thing that your project will make possible.

At the end of the panel (or panels), the program officer will have the ranking board, all the reviews, the panel summaries, and his or her own notes about the discussion of each preproposal. From this, the program director must formulate a recommendation about which preproposals to invite and decline. Obviously, the panel's ranking is a key piece of information, and the program director will often follow the panel's advice closely. However, the program director often will not follow the panel's ranking precisely. This is because the program seeks to balance the portfolio of invitations in several ways: geographically, in terms of beginning investigators, investigators from undergraduate institutions, and so on. Beyond these diversity variables, the program officer will also want to make sure that his or her recommendations cover several aspects of the program's subject area. The program officer may also feel that the panel ranked a few of the preproposals incorrectly. For example, a panel may give undue deference to an investigator's standing in the field, even though a nearly identical proposal from a junior investigator received a low rating.

One final point about overall procedures and policy at NSF: you should try hard not to overreact to snippets of information you hear about NSF in general and your target programs specifically. It often happens that a friend returns from serving on a panel and you hear something like, "NSF just isn't funding basic population ecology anymore." Or a friend simply complains to you, "NSF won't fund two proposals for a single investigator," or "NSF won't support projects with budgets larger than x," or "NSF doesn't like projects on organism x," and on and on and on. You need to treat all information of this type with caution. No statements like these will be correct about all of NSF, which is a very diverse organization, but there may be a grain of truth in claims like these for specific programs. For example, it is usually pretty difficult to get two awards in the same competition. However, with the right timing and projects, it's quite possible to have two or more NSF awards at once. Indeed, a lot of our colleagues make their living this way!

Many of these rumors can be safely ignored, because even if true they won't affect your best practices as an applicant very much. However, if you hear something that you think might affect your submission strategy, you need to confirm it with your program officer. Add it to your list of things to discuss in your presubmission telephone meeting with the program officer.

Commandment #6: Submit two preproposals to maximize the likelihood that you will receive at least one invitation.

How many preproposals should you submit?

You need to choose a submission strategy that matches your funding and research goals. At this writing, the rules allow you to submit two preproposals to IOS and two to DEB. This means that a scientist whose work straddles these two intellectual areas (environmental biology and organismal biology) could in theory submit four preproposals as an applicant. Moreover, there is no limit to the number of preproposals you can participate in as a collaborator. How many should you submit? Given that DEB and IOS now hold only one major competition per year, you should think long and hard about this.

Obviously, there are lots of considerations here. First, do you have four competitive ideas? Do two of them genuinely fit IOS while the other two really fit DEB? If so, how should you manage your time developing them into preproposals? Let's consider two extremes: submitting four proposals versus submitting only one.

Submitting four

If you submit four (two to DEB and two to IOS), you would ideally also submit each of them to separate programs within DEB or IOS. For example, you might submit to the animal behavior and the organism-environment interactions programs in IOS and to the evolutionary ecology and the evolutionary processes programs in DEB. Now you have to be a pretty broad and skillful scientist to do something like this. The reason you might want to spread your preproposals across programs is that NSF considers each program a separate competition, and since program officers are asked to recommend a diverse set of recommendations, you are unlikely to get two invitations from the same program. However, this extreme risk-spreading strategy is fraught with difficulty. Not only is there the very high hurdle of constructing four preproposals that actually fit in four different programs, there is the very real possibility that your strategy will be destroyed by horse trading at NSF. So if you attempt something like this, you must speak with *all* the relevant program officers to make sure they will keep your proposal in their program.

Submitting one

The opposite extreme, of course, is that you submit a single preproposal to a single program. Data from the first year of preproposals suggests that this is what most applicants do (although NSF expects that applicants will start using more sophisticated strategies as they settle into the new system). This clearly has some advantages. Chiefly, you can focus on a single idea and give it your best efforts. Moreover, even if your proposal gets horse-traded, it will still be reviewed somewhere, so this is pretty safe. The disadvantage is that if this lottery ticket isn't a winner, you have to wait a whole year to try again. Moreover, as noted above, if you submit two preproposals to the same program, you are effectively competing with yourself, so the single submission strategy eliminates that concern.

Obviously, there are many options that lie between these two extremes. You should choose your strategy based on an understanding of the rules and procedures and on your funding goals. If you just need some money (for example, because this it what you need to get tenure, or to keep your lab funded), then you should be edging toward a risk-spreading multiple submission strategy; however, if you have one clearly focused project and limited time to prepare and write, then one well-crafted preproposal is probably better than two sloppy ones.

Some generic advice

Many investigators fit in one and only one NSF program, yet the once-per-year cycle represents a significant threat to the continuity of their research program. If this describes you, I recommend that you submit two preproposals to your target program each January. Although you will be competing against yourself, this strategy does increase the chances that you will get at least one invitation. However, to do this you need to choose two topics that meaningfully spread your chances. Program officers recommend that you consider one conservative preproposal that represents a more or less linear progression of your research program and another that is significantly more risky such as one that takes on a big question with a novel approach. I think this is sound advice, but be sure that both of your topics satisfy my criteria for 'a good idea" outlined above.

Commandment #7: Talk to your program officer on the phone. Send him or her a well-crafted summary of your idea prior to this discussion.

Talk to your program officer

Now you've done an award search for your target program, you've thoughtfully developed an idea justified with a strong "what becomes possible" argument, and you've developed an understanding of NSF basic procedures. The next preparatory step is to write a brief summary of your idea (one to one and a half pages). I recommend that you follow the outline for the "overview and significance" section that I give in Part II (page 50). You should expect to spend a few weeks crafting a respectable document, because you want this to spark the program officer's imagination and generate a little enthusiasm. (It's not a waste of time because you will reuse this prose for your preproposal.) Once you have something you're happy with, you will send it via e-mail to your program officer with a request for a telephone meeting. You DO NOT want to restrict this conversation to an e-mail exchange, because a program officer can (and will) tell you much more over the phone than via e-mail. In the text of your e-mail you may say something like:

I'd like to make an appointment to speak with you on the phone to discuss the suitability of my project (description attached) for your program and to clarify some issues about how preproposals are evaluated.

You really want to do this *at least a month* before the deadline, because 1) you need the time to incorporate what you learn in your preproposal, and 2) the program officer will be swamped with late callers in the last two weeks before the deadline.

Although the main order of business in this conversation is to determine whether your project is appropriate for your target program, you will also be working to obtain little nuggets of information. Even if your proposal is technically appropriate, the idea might raise warning flags. Most importantly, even if your project seems to fit exactly with the list of current awards, you want to know if your topic is one the program is enthusiastic about (do they already have enough projects on the topic?).

You should also ask for clarification about any details of the procedure that you don't understand. You might ask about how many invitations the program expects to make. Is the funding for the program expected to increase, decrease, or remain the same? If you anticipate that your full proposal will require a large budget, you could ask whether the program is offering any budgetary guidelines. Ask whether there are any areas the program is encouraging (even if they are not your area): "Are there other programs at NSF that I should consider for this project?" If you're eligible for the CAREER program, you should ask about that and how a CAREER submission might affect your preproposal and eventual full proposal. You can also ask a few questions about the program officer's style and interests: "What do you see as the emerging topics in our field? What sort of proposals would you most like to see?" If you don't already know the program director, it's a good idea to investigate the program director's personal research interests before you speak on the phone.

Even if you violently disagree with NSF's policy or actions about something, you need to keep this conversation friendly and professional. Do not complain, and do not use this as a forum to revisit old wounds. The program officer did not set NSF's overarching policy, and chances are she was not there when your last proposal went down in flames. If a past rejection raises a question about program-level procedures or expertise, be sure to address this in the most positive possible way. You might say, for example, "From my experiences with past submissions, I got the idea that your panels are especially critical on nonexperimental research. Is that a fair conclusion about your program?"

The program officer has the most accurate and up-to-date information about your target program, so you want ask enough questions to let her give you her assessment. Moreover, it won't hurt if you make a favorable impression on the program officer. A program officer won't fund a crappy proposal because she likes you, but it won't hurt if she thinks well of you.

PART II:

WRITING

As you probably know by now, the four-page preproposal is very hard to write well. Basically, you need to achieve in four pages everything you used to do in fifteen, and you need to do it with clarity and scholarship. The first preproposal competition was a wake-up call for many seasoned investigators. They learned that getting an invitation is not a "gimme." You must bring your best game.

You can (and should) find many sources of advice about grant writing and about writing in general. You should use them. Most, but not all, of them contain very helpful nuggets of advice. However, the single most important thing you can do to write a strong preproposal is to do the preparation discussed in part 1. That is (1) develop a strong research idea with a compelling answer to the "what becomes possible question," and (2) understand your target BIO program and its procedures so you can write a proposal that targets the right audience. So if you think you can cheat by skipping right to this "write a preproposal" section, don't. Go back and work through the steps needed to develop a strong idea targeted for the right program.

Of course, you do actually have to write up your hot idea and make decisions about how to present it. Writing well matters. This section offers a few tried and true ideas about grant writing as well as some personal opinion about writing style and philosophy.

> *Commandment #8: Write for your readers: follow a clear plan, use the active voice, and avoid acronyms.*

Reader-centered writing

You can improve your preproposal in many ways, and as I remarked earlier, you'll find many sources of advice about this. The central philosophy that I find most helpful is to remember that your writing must speak clearly and forcefully *to the reader*. In the case of NSF proposals, panelists and program directors are overburdened with reading, so if you are a smart proposer you will do everything you can to make your proposal a pleasure to read. This section offers a few basic guidelines about how you can achieve a reader-centered preproposal.

Writing is work

First, remember that writing well is hard work. It requires solid preparation in the form of a well-developed idea and a clear outline; careful crafting of each sentence and paragraph; and ruthless editing and rewriting. I tend to put people who tell me that writing comes easily in the category of used car salesman and congressmen.

So face up to the fact that writing well is hard work. It follows that you must not underestimate the time required to write a four-page preproposal. Don't fall into the "it's only four pages" trap. It *is* only four pages, but in just four pages you need to make an irresistible case for your idea. You may have also observed that "winning" in this grant system is important to your career. So take the time to get it right. I'd recommend that you allow between one month and six weeks for the writing — that is, after you've developed the main idea. You may need as much a six months if you're still in the idea-development stage.

Have a plan and tell a story

Strunk and White's indispensable book on style states this as a commandment, and that's how you should treat it. Nothing is more frustrating to a panelist than a disorganized proposal. I will give some advice about my favorite "generic grant writing" outline (the so-called Russell system) below, but which outline you use is less important than having a clear outline. Again, take time to develop your outline. Make sure that each paragraph leads the reader logically to the next. Under the "writing is work" headline, remember that your first outline is a draft. Expect to revise it and flesh it out over three to four revisions.

Of course, one of the reasons you need a clear writing plan is because you can't ask the reader's brain to assemble random snippets of your argument for you. You must assemble the argument for the reader. Indeed, data from reading-comprehension studies shows that so-called narrative text is the easiest for readers to understand. Narrative text is writing that tells a story. It has characters, actions, and a plot. You can surely remember the effortless way in which you understood the fairy tales that your parents read to you as a child. You can understand even subtle details of these stories while you're sleepy!

Well, panelists are often sleepy, so tell them a well-crafted story with characters, actions, and plot. I'm not advocating that you write a proposal about trolls spinning straw into gold, but there is much to learn from the ease of comprehension that stories achieve. You need to maintain a high standard of scholarship, and you must not insult the reader's intelligence. You will achieve proposal-writing Zen when the panelists love what you've written but have no inkling that you have told them a story. Having a clear, well-crafted plan to your writing is obviously the first step toward crafting a preproposal that capitalizes on the power of narrative writing.

Lessons from fairy tales

Obviously, a key lesson from storytelling is that you should write simply and clearly. This means short sentences, it means tight paragraphs where each sentence moves the paragraph along, and it means that each paragraph moves the overall narrative to the next stage. In this short section, I offer three dos and don'ts to help you achieve narrative Zen.

Avoid acronyms

This may surprise you, but NSF employees love acronyms even more than most scientists. If you've spent any time trying to figure out NSF programs and the relationship between them, you have had to disentangle a morass of acronyms. The dirty little secret of acronym-lovers everywhere is that acronyms are useful for writers and not for readers.

They do not meaningfully shorten your text, and unless they are of the rare, universally understood type (DNA, NSF, USA), you should really strive to avoid them. Unfamiliar acronyms place a burden on your reader's short-term memory, so they violate the principle of reader-centered writing.

In our labs and research groups, we might find it useful to create acronyms for note-taking and fast communication. In my group, we give each experiment a three-letter code such as SIG or APO that we use internally, on computer files, and so on. You probably do something like this too. If so, you need to remember that these acronyms and abbreviations will confuse people (i.e., panelists!) outside your lab. When in doubt, write it out! *You should never use an internal, research program–specific acronym in an NSF proposal.*

We get into a gray area when we talk about acronyms used by smaller subfields, for example, all the labs that use a procedure they might call PEPQ. If you choose to write PEPQ in your proposal, you need to be confident that all possible panelists will effortlessly understand it. It doesn't really help that much to write out the full name the first time you use it and the acronym thereafter. You're still burdening your readers with a stupid decoding problem when you want them to focus on the importance of your project.

Use the active voice

Using the active voice may seem like tired advice, yet scientists (that's you, dear reader!) still overuse the passive voice. Active-voice sentences tell a simple story with nouns and verbs. They exploit the power of storytelling prose, and panelists will understand your point faster and more clearly if you use the active voice.

If you tend to write in the passive voice, you owe it to yourself as a writer of serious scientific prose to work on this issue. I'm not a purist; the occasional passive-voice sentence can make you sound erudite, but a paragraph filled with passive-voice sentences places a burden on your reader.

I suggest that you do some exercises to learn the active-voice habitat. A tried and true approach is to take a piece of your own prose (two typescript pages will do) and highlight all prepositions and all forms of the verb "to be." Now rewrite this text to completely eliminate the highlighted text. Remember that you can write any sentence in the active voice, so don't wimp out. Eliminate every highlighted word without adding forms of "to be" or prepositions. You may hate some of your new active-voice sentences, but that's not the point here.

After you've done this exercise, you'll find yourself writing active sentences as you draft your prose. You may choke on sentences where you prefer to not identify the actor:

It is widely believed/assumed/accepted that...

The apparatus was constructed by...

I find both of these common formulations odious and lazy, but many investigators I admire simply cannot write without them. Please consider telling the reader *who* built the apparatus and *who* does the believing. Your writing will be clearer if you do. Writers sometimes prefer constructions of the "it is widely believed" form because it lets them make a broad claim about the current state of knowledge without actually defining the scientists who do the believing. I don't think this fools anyone, and panelists will appreciate your precision if you have the guts to put it into the following form:

Many behavioral ecologists believe...

Give specifics first

Scientists value generality, and we often want to make general and abstract claims up front. For example:

Because predators must learn the association between color pattern and consequence, aposematic coloration only protects prey animals in a population of experienced predators. For example, blue jays that have consumed toxic monarch butterflies will avoid them, but naïve blue jays will readily attack monarchs.

Now this is a pretty inoffensive piece of scientific prose, and it is the sort of thing panelists often read as the applicant develops his idea—a general claim followed by a specific example. Now consider the same idea with two ideas expressed in reverse order.

Blue jays that have consumed toxic monarch butterflies will avoid them, but naïve blue jays will readily attack monarchs. When predators must learn color-toxicity associations, aposematic coloration only provides protection when most predators are experienced.

Giving the example first exploits the power of narrative writing because the example is an easily understood ministory, so that when you make your more general claim, the reader's mind frames it in terms of the story you've laid out.

Commandment #9: Don't overestimate the panel's expertise or underestimate the panel's intelligence.

This statement paraphrases a common piece of advice given to young journalists, and it's dead on when it comes to grant writing. You must not overestimate the panel's expertise to the extent that you skip needed general background and belabor excessively technical details. Yet you must also write for an audience of intelligent and generally well-informed colleagues. The key to reader-centered writing is to know and write for your audience, and doing this correctly means that you steer the right course through this difficult terrain.

Sadly, investigators—especially young investigators—commonly overestimate the expertise of the panel. Proposals that make this mistake are written as if they will be read by someone who "wrote the book" on the applicant's topic. Although a preproposal written at this level might demonstrate the applicant's scholarship, it is very likely to fall flat in two ways. First, it will not give enough background for a general reader to understand where the topic fits within the larger field. Second, because it assumes that the readers are "up on the topic," it will often fail to make a strong argument for the general importance of the proposed research. You must write for the general audience of colleagues in your field. Think, for example, about colleagues who teach a survey course in your field but are not experts in your specific topic, but who will know the textbook-level issues that surround your topic.

The opposite mistake — insulting the panelists' intelligence — is less common, but when it happens it usually involves more senior investigators. Some intelligence-insulting errors come off as arrogant attempts to bypass the niceties of science: "Bloggs thinks his fame is a substitute for explaining his experimental design." The most devastating insult to intelligence is when a proposal substitutes some intellectual slight-of-hand for a clear argument about why this research is generally significant. This can take many forms. A proposal might rely too heavily on the applicant's reputation or on trendy topic or analytical technique. If you follow my advice and methodically develop and articulate your "what becomes possible" argument, you will be on safe ground here. Applicants get into trouble when they exaggerate claims about the importance of their research.

How can you know what is common knowledge for your target panel and what is too technical? You need to know your audience, and you have two tools to gain this knowledge. First, you can do an award search to refine your thinking about who "the community" for a given program is. Second, you need outside readers (note the plural!). The right outside readers can tell you when you're assuming too much about their background and when you're bogging them down in technical details.

How should you structure your proposal?

How should your structure your preproposal? Officially, you are free to use the four pages of your proposal in any way you see fit as long as you address both intellectual merit and broader impacts, and follow the GPG guidelines about fonts and margins. IOS and DEB recommend five sections (Conceptual Framework; Rationale and Significance; Hypotheses or Research Questions; Research Design; Broader Impacts). This is just a recommendation, although it represents a common-sense structure.

Recommended outline and space allocation

You will be able to find many sources of advice about how to structure your preproposal, and most of these grant-writing systems have some value. As long as it is logical and orderly, the outline you adopt is less important than the idea you propose. However, the outline you use does matter, and I begin my recommendations by acknowledging that I am very fond of a grant-writing system developed by Stephen Russell and David Morrison. It has served me very well, and my advice is strongly colored by their approach.

Russell and Morrison operate a consulting firm, Grant Central, that conducts grant-writing clinics. Many universities contract with Russell and Morrison to conduct clinics on their campuses. If you have the opportunity to attend one of these, I strongly recommend it. You should go to the Grant Central website and order a copy of their NSF grant-writing guidebook.

Recommended outline

You are free to organize your four pages of text in any way you think best for your projects as long you address both the intellectual merits and the broader impacts as NSF requires.

I recommend, however, the following outline. To begin, I will simply list the four sections I recommend with a brief explanation of each. Then I will give detailed instructions for each section.

Here's what I suggest:

Overview and Objectives (1 page). Write this first. This section must give a broad overview of why your project matters, and in doing so it must generate enthusiasm for your project. It should explicitly state two or three specific aims that will collectively address your objective.

Background (1.25 pages). Write this third. The goal of this section is to provide the details and scholarship needed to connect what you propose to do to the objectives you stated in your "Overview and Objectives" section. It will have three subsections: "Review of Relevant Literature," "Background," and "Preliminary Studies."

Research Design (1.25 pages). Write this second. In this section you will describe what you propose to do. It should consist of one subsection for each of specific aims you listed in your "Overview and Objectives" section. You should strive to keep a balance between the aims, meaning that you should describe each of them in the same way and with the same level of detail.

Broader Impacts (0.5 pages). Write this fourth. You must lay out a compelling program of broader impacts in this section.

My suggestions flow from my experience as a program officer in the first year of the preproposal system. The biggest surprise to both program staff and applicants was that panelists and reviewers still cared about the nitty-gritty of the research even though applicants have only four pages to develop their idea. Applicants received reviewer comments like, "Will this assay work?" or "Can this project succeed in a typical grant period?" In designing the preproposal system, NSF imagined that the preproposal would establish initial interest in the question, whereas full proposals would bring us back to the nitty-gritty question of feasibility and so on. However, panelists didn't drink this particular Kool-Aid; they just can't stop asking themselves, "Can this be done?"

I recommend, therefore, that you keep a small blurb on preliminary studies in your outline as part of a your larger "Background" section. Obviously, you will have to dramatically reduce the detail compared to what you would normally write, but you can convey a great deal about the excitement and the feasibility of your idea with a paragraph and a figure. The "Preliminary Studies" subsection can provide many useful functions: it can generate enthusiasm for your idea ("Hey, look at the cool result we already have!"); it can make the point that the proposed project is feasible in your hands; and it can provide procedural details while simultaneously reducing the need for procedural detail elsewhere by showing that you already have overcome many hurdles by generating something interesting.

I'm belaboring this because NSF and several other sources of advice about preliminary proposals tend to devalue preliminary data. While I agree that you will have to dramatically reduce this to stay within the four pages, I feel that the advantages are too great to forgo this.

Step-by-step recommendations

Writing step 1. Write "Overview and Significance"

Assuming that you have already developed a good idea (or ideas) and identified an appropriate program, I recommend that you begin your preproposal writing a 1.5-page chunk of text along the lines of the "Overview and Objectives" section recommended by Russell and Morrison. This will be longer than you want for the final preproposal, but I recommend that you do this exercise as if you were writing a full proposal for several reasons. First, it will be a useful template for the remainder of your preproposal; you can groom it as appropriate when it comes time to write the preproposal. Second, it will be useful when you do write your invited full proposal. Third, you can (and should) use this as the extended blurb you will use to run your idea past your program officer (or program officers if you plan to approach more than one program). Although you will need to cut this to one page eventually, you should begin by writing this because it the key to generating enthusiasm for your project and because it can serve as a template for the rest of what you say in your preproposal. I give more details about writing this section below, including an extended example.

An example "Overview and Significance" section

As Russell and Morrison assert, this section is absolutely critical for generating enthusiasm and as a guide to the remainder of your preproposal. Given its importance, this booklet devotes considerable space to the problem of writing it well. It's easiest to begin by dissecting an example, so what follows is the "Overview and Significance" section from a (successful) full proposal that my colleague Aimee Dunlap and I wrote. Please read it analytically, considering the content and the underlying mechanisms of the text. Try to notice what it says and what it doesn't say. We'll review the rationale behind each paragraph later.

For students of behavior, learning is basic. Studies of learning cut across disciplines and levels of analysis from Pavlov's seminal descriptions of conditioning to Kandel's molecular analyses of synaptic plasticity. For students of biology, evolution by natural selection is even more basic. Yet, evolutionary ideas play a small role in the larger world of studies of learning dominated by the disciplines of psychology, cognitive science and neuroscience. The difficulty with meaningfully connecting learning and evolution is not that we lack ideas. We have a fairly large family of models exploring the *idea* that statistical properties of the environment can, over evolutionary time, select for or against learning, but we have very little data. Controlling or measuring the theoretically important variables over many generations has seemed like an insurmountable obstacle. We need to test the predicted effects of these statistical variables on learning evolution in order to refine our models. Empirically validated models of the evolution of learning will contribute to an enlarged perspective on learning that will apply the power of evolutionary analyses to this fundamental attribute of behavior.

Our *long-term goal* is to develop a rigorous evolutionary approach to cognitive phenomena that influences the research programs of cognitive scientists and neuroscientists. The *objective of the research proposed here* is to experimentally evaluate hypotheses about the evolution of animal learning abilities. The *central hypothesis* of this research holds that two statistical properties of the environment---reliability and uncertainty---influence selection for learning abilities. In developing this hypothesis, we created a simplified version of existing models based on our experimental preparation. This model considers a two-stage life history in which animals can gain experience in the first stage, and use their experience in the second. The model then compares the fitness consequences of learning (using experience) to 'averaging' (i.e. not using experience but choosing the action that is best 'on average'). This simple model provides a foundation for the further experimental and theoretical analyses of learning and evolution. The *rationale for this proposal* is that although we have several elegant models of the evolution of learning, we have little data to support or contradict them. This proposal will use a recently developed experimental preparation that allows us to directly test these ideas for the first time. We are well-prepared to undertake this research. Our laboratory has successfully completed two studies using this experimental preparation. PI Stephens has published several well-cited theoretical papers on the selective value of learning and information processing. In addition, we have developed significant connections with psychologists, cognitive scientists and neuroscientists that give us an opportunity to frame questions that will influence these fields.

Specific Aims. We will address our objectives via three specific aims:

Aim#1. Fully characterize how 'reliability' and 'uncertainty' interact to influence selection for and against learning. The working hypothesis for this aim is that environmental uncertainty (i.e. change) determines the selective value of learning in combination with the reliability with which experience guides action. This aim will extend the preliminary work of Dunlap and Stephens (2009, Proc. Roy Soc.).

Aim#2. Assess the effects of two types of costs on selection for and against learning. The hypothesized costs of learning take two forms: physiological costs and opportunity costs. The working hypothesis for this aim is that opportunity costs will significantly reduce selection for learning while the other costs of learning will have more subtle effects.

Aim#3. Characterize the interaction between preference and learning. The
working hypothesis for this aim is that unlearned preferences can impede
selection for learning, and learning can impede selection for unlearned
preferences. We will test these ideas by manipulating preexisting
preferences, and exploiting selected lines that differ in learning ability.

The proposed research is innovative in two ways. First, it applies
behavioral ecology's evolutionary perspective to a central problem in
psychology and cognitive science. Second, in contrast to the typical
methods of behavioral ecology, it uses experimental evolution to test an
adaptive hypothesis directly. We expect that Aim#1 will show the
fundamental interaction between stimulus reliability and environmental
uncertainty that determines the selective value of learning. Understanding
the interaction between uncertainty and reliability is critical to interpreting
and predicting adaptive variation in animal learning abilities. Aim#2 will
show how our basic model can be extended to consider the costs of
learning, and it will emphasize the role of opportunity costs that
investigators have often overlooked. Aim#3 will document the dynamic
interaction between learning and unlearned preferences. This idea has a
long history in studies of learning evolution, and this will be the first direct
experimental test of this claim. Taken together the proposed studies will
rigorously demonstrate the action of selection on learning abilities, and this
will help form the foundation for an evolutionary approach to cognitive
phenomena.

Background & Significance

Significance. Learning is universally recognized as a fundamental aspect of
behavior. For example, students of behavior recognize the key
accomplishments in the history of learning studies (e.g. Pavlov, Thorndike,
Watson, Skinner, Hebb, and Kandel) as milestones in behavioral science.
Although, learning is often viewed as a 'psychological' topic, interest in
learning spans the broad disciplines of behavioral science from
neuroscience (e.g. Morris, Kandel & Squire 1988; Kesner & Martinez 2007;
Maia 2009) to ethology (e.g. Thorpe 1943; Lorenz 1981; Gould 1986) and
behavioral ecology (e.g. Shettleworth 1984; Stephens 1991; Dukas 1998). In
2009, investigators published 19,522 papers that mentioned 'learning' in
their titles or abstracts (Source: ISI Web of Knowledge, Thomson Reuters).
Papers on learning evolution form a tiny subset of this larger literature (519
papers in 2009 have learning and evolution in their titles or abstracts, and
in most of these the 'word' evolution refers to something other than
biological evolution). With a few exceptions (e.g. Hollis et al 1997; Dukas
2000; Mery & Kawecki 2002; Mahometa & Domjan 2005) most work on

learning evolution has been theoretical (e.g. Plotkin & Odling-Smee 1979; Stephens 1991, 1993; Papaj 1994; Bergman & Feldman 1995), and this is probably because the theoretically important variables are difficult to measure or control. This absence of hard experimental facts means that non-evolutionary students of learning can easily dismiss ideas about the evolution of learning as interesting but speculative. *The proposed research has intellectual merit because it will provide direct experimental tests of long-standing ideas about the evolution of learning and by doing so it will help to shift learning evolution out of the realm of speculation.* Empirically validated models of the evolution of learning will strengthen the connection between evolutionary biology and the larger community of behavioral scientists. These models and techniques developed to test them could, for example, be used to explore the phenomenon of 'prepared learning' or 'selective associations' (e.g. Seligman 1971; Logue 1979; LoLordo 1979), in which animals learn some associations better than others. In addition, since our studies will produce genetically distinct populations that differ in learning abilities they may open doors to future mechanistic analysis that exploit the techniques of molecular or quantitative genetics.

In the submitted proposal, this piece of text occupied about 1.6 pages (too long for a preproposal). I recommend writing a '"one-and-a-half-pager" like this—with objectives and significance sections—because most to the text will be useful to you as you write your preproposal, even if you will eventually cut some of it or shift it to another location.

Again, this section follows the Russell-Morrison plan as interpreted and implemented by Aimee and me. Let's review the elements of this blurb piece-by-piece to see how they both generate enthusiasm and give an overview of the investigator's plan.

1. First paragraph: the grabber. This paragraph requires a lot of thought and usually many revisions. You must grab the reader's attention ideally by making connections to things that the reader holds to be central or fundamental. Since we were applying to the animal behavior program, we emphasized the idea that learning is a basic process in behavior. After you grab attention, however, you must also establish that there is a significant problem to be solved. In the Russell-Morrison

terminology, you need to convince the reader that there is an important gap to be filled and that filling it will make significant new things possible. I think this paragraph is sacred. Modify the basic structure of this paragraph at your peril.

2. Second paragraph: the details. The basic elements of this paragraph (see the italicized items in the example) are pretty obvious. Notice, however, that they are all interrelated, so you need to be clear about the difference between the objective and central hypothesis, for example. The distinction between the long-term goal and the objective of this proposal is critically important. Stating the long-term goal gives you an opportunity to trot out your powerful "what becomes possible" argument even if you expect that it will take several funding cycles for this research program to reach its full potential. At the same time, it helps make your more modest objective seem important. In its full form (as in our example), this paragraph gives a little sales pitch about the research group. This can be eliminated or dramatically reduced for a preproposal.

3. Third "paragraph": the specific aims. This is a bulleted list of the aims of the projects. In choosing your aims, try to give them roughly equal weight. Ideally, they should represent approximately equal amounts of research activity. In addition, you want to avoid situations where aim number two isn't worth doing if aim number one fails. Each aim should stand on its own intellectually. After you've stated the aim, you can add one or two sentences that state the subhypothesis this aim addresses, the experimental approach, and the expectation, but you should keep it short. In a preproposal, you might simply state each aim if you're strapped for space. Some of my colleagues feel that you don't need a list of aims like this in a preproposal. However, I feel that seeing them all together helps the reader see how they collectively address the objective.

4. Fourth paragraph: expectations. This explains why the project advances the field, addressing each aim separately and finally emphasizing what you expect the collective contribution of the aims will be. The traditional formula is to provide an

expected outcome for each aim—and it is very effective. If space is at a premium, you could reduce this to a single overall expectation. Your claims about what becomes possible would go here, so skip this at your peril.

5. Fifth paragraph: significance. This is where you start showing your scholarship and command of the literature. This paragraph substantiates the claims of the existence of the problem by a very short literature review. In the middle of the paragraph you make a specific credible statement about why this project has intellectual merit. You finish the paragraph by substantiating your claim about your intellectual contribution. In my recommended preproposal outline, this paragraph would go in the "Background" section, but you should write this as one unit because you'll need it when you send this blurb to colleagues and program directors for comments.

That's it. It should take less than two pages. You should expect to spend about two weeks writing a good one. It's worth the effort because it will frame everything else in your preproposal. Notice a few things. First, this design is very cunning because it not only contains many of the things a reviewer wants to know when evaluating a proposal, it also has many elements designed to generate enthusiasm for your idea. Second, notice the way this approach separates the long-term goal of your research program from the short-term objective of this project AND makes both of these explicit! This is critically important to answering the "what becomes possible" question, but it also takes some pressure off "this project" as long this project's contribution to the longer-term research program is specific and credible. Finally, notice what's not here (yet): preliminary data and experimental design for your research aims. This approach defers the nitty-gritty.

Writing step 2. Draft your "Study Design" section

With the objectives clearly laid out, your next step is to write the section on study design. (Notice that this plan means you will write the third section of your preproposal second!) This is where you explain what you're actually going to do.

You have between 1.25 and 1.5 pages to cover this, if you're following my outline. This is one of the places where the four-page limitation is very difficult. You need to at least outline how you intend to achieve the objectives you listed in the "Overview and Objectives" blurb. Evidence from the first year of the preproposal system is that panelists expect to understand how you can achieve what you plan. If it is not already clear to them, you must make it clear.

You should set aside roughly half a page for a subsection devoted to each of the aims you listed in the "Overview and Objectives" section. Use the title you gave the aim in the "Overview and Objectives" section as the subheading for this subsection. This will make your preproposal seem well organized and coherent. Ideally, you should give roughly equal weight to each aim and provide similar information about each of them. You should begin each subsection with a very short paragraph that reminds the reader of the underlying goal, central hypothesis, and expectations for this aim. Then you need to describe what you intend to do. In a preproposal, you will have to make difficult judgments about what you must say and what you can leave out (see the section entitled "How Much Detail?" below). In general, you should keep this at a fairly high conceptual level, because this will generate enthusiasm and keep the reviewer's focus away from his or her procedural hang-ups. You should end each subsection with a summary paragraph with three elements: how you might deal with any anticipated problems; a nonrepetitious reminder of your expectations for this aim; and generalities that remind reviewers of the connection between this aim and the larger things that your research will make possible. The space you devote to these three elements might vary depending on the circumstances. If you know that panelists are likely to obsess over a particular "anticipated problem," you might devote a whole paragraph to this. Otherwise, you could choose to skip this entirely.

In half a page, you can only sketch what you intend to do, but you must be careful to provide information about any aspect of your project that is unfamiliar or questionable to your target panel. Here are some areas where you need to be especially careful:

1) If there are several techniques to achieve the same outcome (for example, measure the relatedness between two individuals, or assay the degree of DNA methylation), then it's likely that some diversity of opinion exists about the advantages and disadvantages of different approaches. In cases like these, you must state a clear rationale for the technique you propose.

2) If you are tempted to drop a few trendy buzzwords into your project narrative, you might be setting yourself up for extra scrutiny unless you carefully define them and explain the role of these trendy ideas in your research. There are always a few hot ideas percolating though our fields. They may be great topics for a research grant, but you must resist the temptation to just toss a few of these hot morsels into your proposal as a cheap way to buy some enthusiasm. It won't work, and it's much more likely to attract negative attention.

3) Beware of "look-and-see" techniques that generate mountains of data. Panels are often skeptical of anything that seems unfocused. An applicant will sometimes suggest a technique, often a '"hot" technique, that will generate an enormous body of data. Genomics and transcriptomics are examples of this type of technique, but there are many others. Techniques like these are great, but using them uncritically and without a clear and focused plan will generate a complaint about a lack of procedural detail.

4) "Standard techniques" won't protect you. In theory, the more familiar panelists are with your proposed techniques, the less you need to say. But it ain't necessarily so. The more the panelists know about your field, the more problems like the first point will crop up.

Ultimately, the only way to know how much detail you need to give is to find outside readers whose expertise parallels what you expect from panelists as closely possible. You probably need to wait until you have a complete draft of your preproposal, because a truly representative outsider reader's advice on just snippets of your preproposal will not be useful.

Writing step 3. Background (review of relevant literature and preliminary data)

Next, you will back up and write the second section of your preproposal. This "Background" section must contain everything a reader needs to know to bridge the gap between your introductory objectives section and your research design section. This includes a very, very brief review of relevant literature, anything you need to say to prepare the readers for your approach (i.e., introducing your study system and the rationale for choosing it), and a tasty hors d'oeuvre of preliminary data that will make a strong case for the feasibility and importance of your project. You need to fit this into about 1.25 pages. It makes sense to write this after you've written your research plan, because once you know what you want to do, you know where you must lead the reader. Since this section needs to be concise, I recommend that you take an analytical approach to deciding what you must discuss and what you can omit. You'll want to go through your draft of the research design section with a highlighter to mark every point that will need support in the background or preliminary studies section. I see this section as divided into three subsections: literature review, required general background, and preliminary data. They hang together, obviously, because they must all be sharply focused on preparing the reader to accept and understand the approach you will propose in the section on research design.

Literature review

The subsection of literature review should 1) demonstrate your scholarship, 2) support your claims about the significance and importance of your project, and 3) prepare the reader for what you propose. You don't have much space for this, so you must have a laser-like focus on these three issues. You cannot offer an open-ended literature review of your research area. Don't even try.

You can steal the "significance" paragraph of your one-and-a-half-pager for this, but you may need to trim it and add a few additional "scholarship-demonstrating" references to make this fill the bill.

General background

This is where you tell the reader what he or she needs to know to understand your approach. If you're studying a butterfly nervous system, then this is where you must tell the intelligent (but nonexpert) reader what she needs to know about butterfly brains in order to understand the studies you propose. This section is elastic. Investigators using very standard techniques in a well-established system may have very little to say here (but beware of misjudging the panel's expertise). On the other hand, investigators with very unusual approaches or study systems may need to devote a lot of space to this. Biologists, as a rule, love a well-chosen, unique study system, so don't be afraid to highlight the creativity of your choices here. This can really help build enthusiasm. (Avoid "stamp-collecting" arguments here, though; no one cares that the gee-whiz bird has never been studied; we care that a study of the gee-whiz bird has the potential to make new science possible.)

Preliminary studies

The subsection on preliminary studies is your opportunity to sell the potential of your project. If possible, limit this to a few sentences in the text, an accompanying data figure, and detailed caption (set in a small but legible font) that gives statistical conclusions and procedural details. Ideally, the supporting sentences in the text will only need to walk the reader though the figure. Required background would contain any details of your approach (such as an introduction to your study system) that your audience does not already know and that they must know in order to understand your study design.

The biggest potential pitfall in the background section is too much detail. You need just enough to support your plans. You do not need to review your field here.

Writing step 4. Broader impacts

The last thing you need to write is the "Broader Impacts" section. Although you write it last, you should have thought about how to integrate broader impacts into your project from a very early stage. As with the other sections of the preproposal, rule number one here is to have a solid program of broader impacts to propose. I outline the key considerations for strong program of broader impacts in part I (page 12). You should not propose things that everyone already does — for example, training graduate students and publishing papers — as your broader impacts. You should develop a plausible program that touches on at least two of NSF's societal goals. In the ideal situation, your broader-impacts activities will blend seamlessly with your research (although this isn't always possible). You should not, however, write a series of vacuous sentences explaining how your project "ticks" all of NSF's broader-impact boxes.

Writing step 6. Putting it all together

Now that you have the pieces, put them together and edit them. Your first job is to make sure the four pieces read as unitary whole. You should revise to ensure that the draft flows logically from one section to the next and that your terminology is consistent across the sections. Don't worry too much about overall length at this stage. If you've kept your sections close to the recommended lengths (yes, I know they will be too long), you should be within a page or so of the four-page maximum. After you're happy with the logic and readability of the whole, you can begin to trim things.

Writing step 7. Assemble the final pieces while your outside readers read

Now you're ready to send your four pages of text to outside readers. Do not skip this. It's is the only objective tool you have to ascertain whether you've got the level of detail right.

While your kind colleagues are reading your draft, you can be working on the other components of your application (project summary, personnel list, COI spreadsheet, and biographical sketches). The biggest job is the one-page project summary. On FastLane, you must now enter this as three separate pieces: overview, intellectual merit, and broader impacts. You can crib most of this from what you've already written. For example, you can draft a powerful overview section from your grabber paragraph, the list of aims, and final sentence about your overall expectations. Similarly, you can build a concise intellectual merits section from your statements about long-term goals, objectives, and significance statements. Remember, of course, that your project has intellectual merit because it has the potential to make new science possible. Finally, the project summary's broader impacts section could be an introductory sentence ("The broader impacts of this project will address societal goals A, B, and C..."), a list of activities, and a final sentence expressing the general significance of these activities. I don't mean to trivialize the summary. It's a very important document. You *can* assemble the draft from the pieces you already have, but you must make sure that it reads smoothly and makes a strong case for your idea.

Once you've written the summary, you have a series of personnel-related jobs to attend to you. You will also need to construct a list of senior personnel. This is very simple for a single investigator, but it can be a significant undertaking for larger collaborative proposals. While you are doing this, you can be assembling the information you will need for the COI document (which takes the form of an Excel spreadsheet). The final nicety to attend to is the biographical sketch(es). Be sure that it or they follow the new overall rules (we have "products" now, not publications) and the special rules for preproposal (information about collaborations should not be present).

Writing step 8. Revise in response to your outside readers

With the help of your colleagues, you will have a chance to anticipate concerns about the level of detail and the persuasiveness of your arguments. Consider their concerns carefully, and revise as appropriate. In my experience there is seldom time (or patience) for two rounds of review, but I have often asked my reviewers for clarification or advice. Occasionally, this back and forth has radically changed my thinking about the best way to revise.

Commandment #10: Use outside readers, use outside readers, use outside readers!

The final substantive step in revising your preproposal is to get feedback from outside readers. This is always a good idea, but the preproposal's brevity makes it critical. With only four pages, it's easy to skip a critical detail or make the level too high or too low. Obviously, you want readers who closely resemble panelists. You might use one reader who is an expert in your subfield, but people like this are unlikely to be preproposal panelists. You really want readers who are your target program's "everyman." If you use one "expert," try to find at least two everyman-type readers who can give sensible panel-like advice.

It follows that you must finish your preproposal in time to allow for outside reading. You should think in terms of finishing ten days before the deadline so you can give your readers enough time to thoughtfully consider your draft and give yourself a few days to incorporate their comments.

I've offered a lot of advice about preproposal writing here, but I consider "use outside readers" to be the single most important thing you can do to improve your chances of obtaining an invitation. Allow time for it, and do it. It will help.

PART III: AFTER THE INVITATION DECISION

NSF will inform the applicants who submitted their prepoposals in mid-January by the end of the following May. NSF has been releasing the "invite" decisions about two weeks before the "decline" decisions with the intention of giving successful applicants an additional two weeks to write. Regardless of whether NSF declines or invites your prepoposal, you have some decisions to make. This section offers some advice about how to proceed. First, I will explain how to read and understand the feedback you get from NSF. The basic issues here are the same regardless of whether you received an invitation or a declination. Second, I will talk about resubmission strategy for applicants whose prepoposal was not invited, emphasizing common reasons for failure and how to correct them. Finally, I will offer advice about how to transform your four-page prepoposal into an effective fifteen-page full proposal.

How to read your reviews

Commandment #11: Focus on the panel summary when you read your reviews, and don't be distracted by "things that piss you off."

You will receive the official decision about your preproposal in an e-mail from the program director assigned to your preproposal. These mass-generated e-mails represent the culmination of the preproposal process as well as an enormous amount of effort for the wizards behind the NSF curtain. You will receive "the e-mail" after the division director formally accepts the program director's recommendation. In making this decision, the division director will review internal documents (the so-called review analysis) that present the program director's rationale for inviting or declining your proposal.

Your reading list

As soon as the decision is released to you, you will have access via FastLane to:

Context statement

This statement is prepared by program staff and approved by program directors. It reads like boilerplate (and a lot of it is), but there are some useful nuggets here, such as how many preproposals were evaluated in this round and how many were placed in each category. It will not tell you how many panels were used in a given competition or who served on the panel. Everyone who applied to your target program sees the same context statement.

Your reviews

Usually, you receive three reviews from panelists (remember, there are no ad hoc reviews for preproposals). Each review has (i) a score on NSF's "excellent, very good, good, fair, poor" scale; (ii) an overview; (iii) a section on intellectual merit divided into strengths and weaknesses; (iv) a section on broader impacts divided into strengths and weaknesses; and (v) a summary statement.

Panel summary

This gives your overall placement on the board (category names vary, but terms like high, medium, low, noncompetitive are the norm). Otherwise the form is similar to a review, but it summarizes the panel's rationale for its ranking of your proposal. This is the most important thing for you to read and understand. It can be frustrating because it is often telegraphic.

Program director/officer comments

This section is often blank, but if something is there you MUST pay attention to it. This field exists to convey messages that are missing from the panel summary. For example, the program officer may tell you that you shouldn't submit this again, or impart some programmatic information (such as, we already have a ton of awards on this topic, so you should work on something else), or point out the program's central consideration in making its recommendation (such as, the program felt the experimental design for aim number one was fundamentally flawed — ouch!).

You will not see the internal "review analysis" of your proposal, which contains the program director's rationale for his or her recommendation and the identities of the panelists. NSF considers this information confidential.

Focus on the panel summary

Although all of this is worth reading, the panel summary is the main thing that matters to you. For example, if a program officer were to someday complain that your revision was not responsive to your reviews, this would almost always mean that you ignored a point in the panel summary, not something in one of the individual reviews. When you formulate your plans for a revision, you really must deal with the criticisms in the panel summary in some way. I find it useful to break down the issues raised in my panel summaries in a bulleted list or outline, so that I can brainstorm "action steps" under each critique. I will say more about responding to criticisms in the separate sections on "invite" and "noninvite" decisions.

Don't overreact to "off the wall" comments

Obviously, it's useful to read the individual reviews because they sometimes give insights into how key points of the panel summary arose. You do not have to respond to each point raised by an individual reviewer (and in many cases you should not). The individual reviews were written before the panel met, and they are the most variable part of the process. It's not unusual for a panelist to do a complete turnaround after the panel discussion, and this is why the panel summary is so much more important. Moreover, individual reviews often say stupid things that make investigators see red. If these "things that piss you off" (TTPYO) are not repeated in the panel summary, you can probably ignore them. My experience as a program officer, applicant, and colleague is that applicants can waste huge amounts of time and emotional energy agonizing over TTYPOs. Getting worked up about TTYPOs will not get you a grant. Simply make a note of these complaints, make a simple plan to make sure that future readers don't come to the same erroneous conclusion, and get back to addressing the issues in the panel summary.

Commandment #12: Take responsibility for problems in your preproposal.

Let's be honest: panelists, reviewers, and program directors are idiots, and they shouldn't be trusted to vacuum the carpets at 4201 Wilson Avenue. A surprising number of seemingly rational colleagues respond to bad news from NSF by blaming NSF or some aspect of NSF. Of course, NSF reviewers and program directors make mistakes, but you will not succeed by dwelling on these mistakes. Successful applicants take responsibility for bad reviews and decisions that go against them. You need to think "what can I do to address this concern," not "how can those idiots think that." This is especially important when you deal with program officers and NSF staff in general. If you harangue them about how reviewers treated you shabbily, you will not get the advice you need to avoid these misunderstandings in your future proposals.

Speak to your program officer for clarification

After you have digested your reviews and the other information NSF has for you, it's time to speak with your program officer for clarification. Your assigned program officer will be indicated in your reviews and in all the notifications (invitations or declinations) that you receive. As before, you should set up a telephone appointment and give the program officer some indication of what you want to talk about. The key point of this conversation is to make sure that you have correctly understood the main points raised by the review process and that your general thoughts about how to respond are moving in the right direction. The program officer cannot give you detailed advice about how to write your proposal. She can't say, for example, that proposed experiment A is better than proposed experiment B. Instead, she can help you decode what the panel summary means and what the panel and program are looking for in a resubmission.

I suggest that you make a list of issues raised by individual reviewers and in the panel summary. Review this list with your program officer. Again, you should focus on the panel summary, and you should keep this list at fairly high conceptual level. You don't want to waste the program officer's time with a trivial question about the use of semicolons, for example. You should ask separately about the concerns of the panel—that is, what does the panel want to see in a resubmission?—and about the program's concerns about your project—that is, what are the most important issues from the program's perspective?

You don't need to include a "response to reviewers"

Regardless of whether your preproposal was invited or declined you will be revising. You'll either be revising and enlarging it into a full proposal or revising the preproposal for resubmission next January. Before I discuss the specific strategies for these two types of resubmission, I want to make a general point about how to respond to reviews. Some grant-giving agencies expect you to dedicate a section of your resubmission to a formal "response to reviews." NSF does not require this in general, although specific programs or solicitations may. Although circumstances clearly vary, it isn't always necessary or even advisable to include such a section. Most of the time, it's better to simply correct the problem and say nothing more about it. If the panel says, for example, that it is concerned about the power of a proposed statistical analysis, it's usually better to simply include a rigorous power analysis in your resubmission and to set aside text that says something like, "Previous reviewers questioned the statistical power of our…In response, we now offer a full analysis of this test and propose an increased sample size of…" This just wastes space and draws renewed attention the to problem that your next reviewers may not care about. Don't waste time on this—just fix it. In my view, this is the best way to proceed 60 to 70 percent of the time. In a few cases, you may need to emphasize the point that you have been responsive to the reviews. This could happen if previous reviewers egregiously misunderstood something, and you may need to politely and constructively "call them out." I would do this only very reluctantly, however.

Resubmitting a declined preproposal

Obviously, you are going to resubmit regardless of whether you received an "invitation" or a "not invite" decision. At this point, however, the strategies for resubmitting your idea as a full proposal and resubmitting it as a preproposal are pretty different, and I deal with them separately. I consider the strategy for resubmitting a preproposal first.

Analyzing what went wrong

If NSF didn't invite your submission, your first job is to decide whether it's worth your time and effort to revise your preproposal. The alternative, which is often a good one, is to start fresh with a new idea (or ideas, since you're allowed two preproposals). I think there are two things to consider here: 1) your reviews (Are they encouraging? Can you address the stated concerns?); and 2) your personal assessment of whether you can take this idea and revise it into your "best shot" at an invitation in the next round.

Your reviews

I don't think it's reasonable to give a hard and fast rule here, but if your panel ranking is in one of the bottom two categories (usually low priority and noncompetitive), you should think pretty hard before resubmitting this idea. At the other extreme, if the panel placed your preproposal in "high" and you didn't get an invitation, you probably have a problem that flows from the program's concerns about portfolio balance (too many funded projects on your topic) or possibly too many preproposals on related topics in the pool during this competition. If this is the case, you really need to talk to your program officer for guidance. The ideal "encouraging enough to resubmit" case is just out of the money somewhere in the second category (usually called something like "medium priority"; in case you missed this, the terms for panel rankings should be in the "context statement" that is available to you via FastLane as soon as NSF releases your reviews to you).

Did you fail the "what becomes possible" test? Regardless of your reviews and ranking, you didn't get an invitation. Your panel summary and reviews should identify specific problems, but you need to be aware that you probably have a problem that isn't written down in the panel summary. You didn't excite the panel! Most preproposals can "walk and chew gum"; that is, they are technically competent and reasonably scholarly. These technically competent proposals really fail because they didn't make an exciting case for the importance of what they propose. When this happens — and it's very common — the issues raised by the panel are really excuses to give your unexciting proposal a lower ranking. As I explained in part 1, the key to making a preproposal exciting is to have a compelling argument about "what your project makes possible."

Principal investigators sometimes have the idea that if they competently address the issues raised in their reviews, NSF somehow owes them an award. While submissions to journals sometimes work like this, this sort of thinking is completely alien at NSF. So address the reviewer's concerns by all means, but also think hard about articulating a compelling argument that your project would advance science by making something new possible.

Commandment #13: Be sure you're addressing the real problems when you revise.

Common criticisms and flaws

I cannot pretend to anticipate all the complaints that reviewers could level against a preproposal. The topics covered by IOS and DEB are far too diverse for me to generate a catalog of possible complaints. Yet several types of concerns do routinely come up. The list below identifies some of these, decodes the implicit NSF-speak where necessary, and offers some advice about how you can address them in a revision.

Unimportant question

This can come up in many guises. The panel summary would often say something such as "the panel was not convinced of the significance or intellectual merit of the problem." This could mean several things, but the most likely possibility is that you didn't know your audience and therefore made a weak "what becomes possible" argument. At a minimum, you need to strengthen this argument (review the discussion of commandment#3 in Part I). You might be well advised to consider submitting this to a different program because this complaint could arise if your idea seems too far afield to the panel. If you suspect this, you should do some snooping about this panel (e.g., an award search, speaking to the program officer, etc.).

Important question/bad design

This type of critique often arises in crowded fields. It often means that investigators in your field agree that this is an important problem (and the panel probably saw several proposals attacking this problem). Obviously, you want to ask yourself if you can address the panel's concern about your approach, but you might also want to consider another topic and perhaps a second preproposal that spreads your risk to another less crowded question.

Too incremental/me too/stamp collecting

I discussed this flaw in part 1. You should watch for this message in the program officer's comments. It's a big red flag, and you need to be thinking hard about a new approach.

Technically unsophisticated

The panel could say this in several different ways, but basically the point here is that your scholarship isn't up to scratch. You may have missed some key bit of background or proposed a study without acknowledging a key difficulty.

Poorly written

Every panel sees a handful of proposals written by smart investigators who just didn't take the time to construct a logical and competent narrative. Misspellings, bad grammar, and other signs of editorial sloppiness will anger panelists. Remember, they have committed a big chunk of time to review your ideas. If you give them the feeling that you've wasted their time, you've lost the battle. The fix is straightforward. Apologize and do the work next time.

The panel wasn't convinced it could be done

Although NSF has tried to de-emphasize questions about feasibility in the preproposal system, panelists still care about them. This is tough because you have only four pages. You need to look closely at the panel's critique and devise a plan to address it in a paragraph of a few sentences. Preliminary data can be helpful, especially if you can put some methodological detail in a nine-point figure caption. You really need outside readers to address this. Only a well-chosen set of readers can help you decide how much detail is too little.

Preproposals often languish in the middle range of "the board" simply because no one is enthusiastic about them. These are often technically competent preproposals. They just didn't excite anyone. This is very common. If you feel that the panel nitpicked your preproposal, this is often the explanation. This can be a terrible trap for the applicant because he or she believes that addressing the nitpicky complaints will turn the tide, but it won't. In my experience this is so common for rejected proposals that you can safely assume it applies to you. You need to think hard about how to generate more enthusiasm in your revision. Strengthening your "what becomes possible" argument is a good place to start.

Strategies for a preproposal-based full proposal

Congratulations, you got an invitation. Now you need to figure out how to translate your four-page preproposal into a fifteen-page full proposal. This section discusses this translation. It is a short section because the story here isn't new. You write a full proposal now very much as you did before, and some dynamite advice exists about how to do this in Russell and Morrison's guidebook. I will, however, try to highlight some of the key points.

The sameness rule

The one basic rule here is that you must propose the same project in your full proposal that you described in your preproposal. This rule of sameness has two components. First, you can't change the senior personnel (e.g principal investigators, co-prinicipal investigators, and senior personnel) without obtaining permission. Second, you can't change the scope of the proposed work. In the simplest cases, the meaning of "changing the scope" is clear-cut. You can't submit a full proposal on animal behavior if NSF invited you to submit a proposal on ecosystem function. In practice, I think you're pretty unlikely to propose a dramatic change in scope. It does seem likely, however, that new experiments or approaches will occur to you as you prepare your proposal. In most cases these will be within the original scope because they will use a similar study system to test similar hypotheses. You should speak with your program officer, however, if you anticipate making a dramatic change of some type.

Writing your full proposal

In making the transition for your successful preproposal to a competitive full proposal, you need to do two things that are a little different from writing a full proposal from the ground up. First, you have reviews from the preproposal round of the process, and you should obviously use these to strengthen your full proposal. Second, you need to enlarge the story you told in four pages to fifteen, so you will need to expend some energy in considering how your full proposal emerges from your preproposal. I discuss these two issues in the next few paragraphs, discussing the "enlargement" question first, because it helps to frame the question of how you should deal with reviews.

Enlarge your preproposal by shortening it first

If you follow the Russell and Morrison system, your full proposal starts with a 1.5-page "Overview and Objectives" section that forms the road map for the rest of your proposal. This is solid advice for two reasons. First, careful thought put into crafting this short blurb will clarify your thinking about what you want to say in the remainder of the proposal. That is, this blurb will identify themes that you can build upon and reemphasize as you write your fifteen-page proposal. Second, if well written, this section is a powerful tool to build the panel's enthusiasm for your idea. My first recommendation about lengthening your preproposal is to begin by shortening it! Look at your four pages and distill it down to 1.5 powerful pages of scientific prose. In broad outline (see page 50), these 1.5 pages should have i) a grabber paragraph that establishes the problem; ii) an objective paragraph that sets the long-term goal, objective hypothesis, etc.; iii) an explicit list of "specific aims"; iv) a paragraph that emphasizes the expected benefits of your research; and v) a "significance paragraph" that provides scholarly documentation of what you have said.

Most of this prose will exist—at least in skeletal form—in your submitted preproposal, so you should be able to slap together a draft of this "Overview and Objectives" section fairly quickly by cutting and pasting paragraphs from your preproposal. Obviously, you will have to revise this to make sure that it reads smoothly and to ensure that it establishes a solid foundation for the 13.5 pages you still have to write. Moreover, you will obviously want to read this with your reviews in mind. You might want to use the strengths identified by the panel to sharpen your statements about significance and intellectual merit. Similarly, you will want to guide the reader away from the panel's concerns about weaknesses.

Once you have condensed your four-page preproposal into a dynamite "Overview and Objectives" section, you can simply follow the Russell and Morrison recipe for constructing a fifteen-page proposal. The only wrinkle is that unlike the vanilla Russell and Morrison system, you have reviews of your preproposal and I will discuss how to address these reviews below.

Dealing with reviews

Since NSF invited your full proposal, it's a reasonable bet that the criticisms leveled against your preproposal are relatively minor. If this is the case, then your job is relatively straightforward. You should address every criticism listed in your panel summary, but this doesn't mean you explicitly identify a paragraph as a "response to previous review." As I explained above, this is usually counterproductive. You want to address each issue without drawing attention to it as a previous critique. If, for example, the preproposal panel expressed a concern about whether you planned the proper control treatment, then you should take pains to describe a control that addresses their concerns in your section of "study design," but you would not say "because previous reviewer complained about the lack detail about controls..." In other cases you might revise a paragraph of your background to show that you do indeed know something about the alternative techniques that reviewers said you were ignoring, and so on. Rinse and repeat until you have addressed all the complaints made by the preproposal panel.

This advice is reasonable assuming that the complaints are minor and nitpicky. If your reviews make more substantive criticisms (for example, by questioning the importance of one of your proposed aims), you will need to reorganize your proposal more profoundly. In my view, you should do this when you condense your preproposal into the 1.5-page overview section that provides the basic plan for your full proposal. If you need to change an aim, then you may also need to tweak the objectives, long-term goals, and so on, because it's important that these introductory pieces form a coherent whole. You will need to be careful not to violate the "sameness rule" if you change an aim or the objective. It is a good idea to speak the program officer about this. If you are responding to a complaint raised by the review panel, it is very unlikely to object to the change you propose. Do not overreact, however. If your preproposal was invited, you have NSF's approval of your idea at least in broad overview, so stick with the broad overview you proposed.

Alternatives to the preproposal system

Conventional grants are for suckers?

Commandment #14: Learn about and exploit alternative grant mechanisms at NSF.

One of most surprising personal discoveries I made during my time at NSF is that several prominent and successful investigators hardly ever apply for normal grants. Instead, these investigators heavily exploit NSF's special programs. They make a career out of visiting the NSF website for announcements, phoning program officers to ferret out the details of new programs, and remaking their research programs to suit different NSF programs. Yes, there's something a little smarmy about this, but the facts are that conventional grant programs are very competitive and funding for them is constantly under threat. In contrast, many special programs at NSF are underused and have only internal reviews, and some of these programs make million-dollar awards. So while I don't really think that conventional grants are for suckers, I do think that *restricting* your applications to conventional grants is a sucker's play. If extramural funding is central to your success as a scientist, you should at least pay attention to these alternative sources of funding.

These programs are not "free money." You will have to work, and possibly work quite hard, to tap into many of these programs. However, they represent another way forward for investigators seeking extramural funding. The investigators who exploit them most successfully are those who have flexible research programs. Your success with these programs will depend on your ability and willingness to write a proposal that fits the program and not on your ability to find a program that finds what you already want to do. Flexibility and creativity is the key to success here. To help you achieve this success, I offer a short list of current "alternative funding mechanisms" at NSF.

Program versus "special" funds

Some of NSF's special programs are "fake" programs in the sense that they do not have any money attached to them. The CAREER program is an example of this. If a program makes an CAREER award, the money for the award comes the same pot of money that the program officer uses to make conventional awards. These "from-program-funds" mechanisms are things that NSF wants to encourage and track, but they are in practice "less special" than the special programs with actual cash behind them. Other programs, like INSPIRE, have a special pot of money set aside for them. You should know about this difference because it affects your interaction with the program officer about these programs. If the money comes from program funds — that is, the program officer's budget — then the program officer is guarding his or her programmatic nest egg when you call to discuss a possible award. If, instead, NSF has set aside money for the program you're interested in, you and the program office are more like coconspirators, because if you have a good idea the program officer can use your research to get more resources for his or her program.

Programs you should know about

CAREER

Many readers will already know about the CAREER competition. However, if you are eligible (recently hired into a tenure-track but untenured position), you've got to take this one seriously. Funding for CAREER comes from program funds and is fairly competitive, but the payoff can be significant. My program typically awarded only one per year. If you're eligible, a CAREER submission gives you a way to circumvent the once-per-year cycle. In particular, if you submit a preproposal in January and you're CAREER eligible, you have more options than others. You can use the feedback from an uninvited preproposal to craft a strong CAREER proposal. If you're lucky enough to be invited, you can submit a CAREER and a full proposal in the summer.

You can find CAREER solicitation at

http://www.nsf.gov/funding/pgm_summ.jsp?pims_id=503214.

RCN: Research Coordination Networks

This is an underused NSF funding mechanism that is intended to provide resources to develop new collaborations. Often an RCN will propose working groups, meetings, and various kinds of organizing activities. The stated goal is to develop new directions in a field (so answer that "what becomes possible" argument!). According to the solicitation, "RCNs do not support primary research." Instead, the idea is that they support activities that should catalyze research via new interactions. NSF recognizes two subthemes, one focused on sustainability and the other on undergraduate education in biology. Most of the RCNs in BIO are of this second type. RCNs are funded via program funds. Each program can determine its own deadline for RCNs. IOS, for example, considers RCN proposals at the same time it considers full proposals. It follows that if NSF declines your proposal in May, you could still put together an RCN proposal for the August deadline.

You can find the RCN solicitation at

http://www.nsf.gov/funding/pgm_summ.jsp?pims_id=11691
.

EAGER

EAGERs (early concept grants for exploratory research — who makes up these acronyms?) are smallish grants for high-risk but potentially transformative research. NSF wants to fund cutting-edge projects that make new things possible, and sometimes the ordinary grant mechanism is just too cumbersome for this. NSF has tried to address this need via several different mechanisms (SGERs, if you're old enough to remember those). The EAGER mechanism represents NSF's latest attempt to meet this need. My experience is that a plausible EAGER needs to hit two basic points. First, you need to make a convincing case that your idea has the potential to make a significant impact on its field (challenge current thinking, offer a technique that makes new questions possible, etc.). Second, you need to make a case this is project would not succeed as a standard program grant. A good argument here is lack of pilot data; if x is true, it changes everything, but it needs to be tested to find out whether it's worth pursuing further, etc. It's critically important that you develop your EAGER application in close consultation with your program officer. EAGERs come from program funds, and they are normally reviewed internally.

Here's a tip for success. Don't ask yourself, "How I can browbeat my program officer into giving my an EAGER for the project I want to do?" Instead, ask, "What component of the research I want to do can be recast as EAGER-appropriate?"

That means potentially transformative and too risky for a normal grant!

The EAGER mechanism is foundation-wide, and the GPG describes this basic rules:

http://www.nsf.gov/pubs/policydocs/pappguide/nsf13001/gpg_2.jsp#IID2.

RAPIDs (grants for rapid response research)

RAPIDs are the less-used cousins of the EAGER mechanism. RAPID projects have some externally imposed urgency; for example, a natural disaster creates an opportunity for data collection, or a set of samples that will only be available for a short time. Opportunities arise more frequently in field-oriented disciplines, but you should not close your eyes to this possibility if you're one of the lab folk. An amusing observation about RAPIDs is that RAPID applications are fairly rare, and a consequence of this is that program officers like to tick "award a RAPID" off their "life lists," especially if they are in a lab-oriented field where external events are rare. Wait for the right opportunity.

As with the EAGER mechanism, the RAPID mechanism is foundation-wide, and the GPG describes the basic rules. Again, it's critically important to get the support and advice of the appropriate program officer. RAPIDs come from program funds. Information can be found at

http://www.nsf.gov/pubs/policydocs/pappguide/nsf13001/gpg_2.jsp#IID1.

OPUS

The OPUS (opportunities for promoting understanding through synthesis) program is a CAREER program for old farts — oops, I meant senior investigators. That is, it provides support for more advanced investigators to complete synthetic work. First, this is a DEB program, so you must somehow fit under the umbrella of environmental biology to apply (although this is a fairly broad umbrella, it will rule out some investigators in BIO). In addition, the applicant must present evidence of a program of publications over a significant period of time that will somehow be synthesized. Finally, the budgeting is unusual because it primarily supports salary replacement for the synthesizing investigator—read "time off to write." Visit http://www.nsf.gov/publications/pub_summ.jsp?WT.z_pims _id=13403&ods_key=nsf12506.

INSPIRE

The INSPIRE program seeks to provide a large chunk of support for interdisciplinary and potentially transformative projects. The minimum award is $1 million, and applications are reviewed internally. Applicants must submit a letter of intent, and then program directors from at least two "academically distinct" components of the foundation must support this letter of intent. The letters of intent are reviewed, and full proposals are invited. In the past, invited letters of intent had a very good chance of ultimate funding, but this seems to be less true as the program becomes more competitive. Nonetheless, it represents a significant shot at biggish chunk of money.

The programs supporting an INSPIRE award must put up a small amount of money, but most of the award comes from foundation-level funds. This means that your program director may see this as a way to leverage program funds, and as I remarked above, this can change the nature of your interaction with your program officer. The INSPIRE program is just finishing its second year (it was called the CREATIV program in year one). It is the brainchild of Subra Suresh (recently departed Director of NSF). Dr. Suresh saw this program with its requirement of interdisciplinary support as a mechanism to support integrated approaches to large problems and address the criticism that NSF is too conservative. So swing for the fences when you apply to the INSPIRE program. I've described the plain vanilla INSPIRE program (called "Track 1" now), but it now consists of three tracks. Track 2 is for larger projects requesting up to $3 million over five years. Track 3 will recognize an especially meritorious subset of Track 1 submissions with a little extra money and the honor of receiving a "Director's INSPIRE Award."

Every investigator should take this program seriously. The key pieces of advice are two. First, work closely with your target programs to craft a letter of intent that proposes a meaningful interdisciplinary project. Many ideas founder because their attempts to combine disciplines are too shallow. Second, be as bold and creative as you can possibly be. Have a strong and compelling argument from "what becomes possible" and how this new possibility will influence multiple disciplines. Information is to be found at

http://www.nsf.gov/publications/pub_summ.jsp?ods_key=n sf13518.

Note added in proof: In response to the federal budget sequestration, NSF has suspended the INSPIRE program for the coming fiscal year (2014). In theory, it will return sometime after congress resolves the current budget morass. Keep an eye on this program for fiscal 2015, if the window re-opens the timeline may be short.

International programs

NSF has an Office of International Science and Engineering that, to be honest, has been in a state of flux recently. In theory, NSF has international programs that are designed to promote scientific exchange with investigators beyond the borders of the United States, but my experience with them is hit and miss. Most of the international programs come in the form of supplemental funding. For example, suppose you have a project in which you propose to collaborate with a Chinese investigator. You should visit the International Science and Engineering website at

http://www.nsf.gov/od/iia/ise/index.jsp.

While there you should check for two things. First, are there any special programs advertised for China (or your target country)? Second, who is the regional program officer? You should then contact the appropriate program officer with a brief description of your project and seek advice about the possibility of funding. Normally, you would submit your project via one of the NSF's normal mechanisms. The Office of International Science and Engineering will cover a portion of the award if the subject-area review leads to an award recommendation. Your subject-area program staff will be very much on your side in trying to leverage money from international programs. In theory, this could push you over the edge for funding if your proposal is eligible for an international supplement, but in practice I've found this effect to be pretty small.

The Office of International Science and Engineering sometimes negotiates agreements with the science agencies of other countries, and this can mean that they will emphasize interactions with some countries over others; so it's worth knowing if your collaborator is in one of the "hot" countries (check the website and consult with the regional program officer). In addition, the office is focused on developing international collaborations rather than maintaining them. So if you have long track record of collaboration with your Chinese colleague, this could actually count against you.

Supplements

Supplements differ from the other funding mechanisms I've discussed because you can't get a supplement unless you already have an award from NSF. Yet supplements are complicated and misunderstood, so they should be a part of your total NSF game. Within NSF's award system, supplements are a general mechanism that NSF uses to encourage certain activities. They exist in two broad forms: the "R" supplements such as the REU and RAHSS supplements that I will explain below, and so-called general supplements, which in the words of a former NSF colleague are "primarily used by former program officers."

The rogue's gallery of supplemental funding mechanisms

REU (Research Experiences for Undergraduates) supplements

These are supplements that support undergraduate involvement in your project, and many investigators will already know about them. They are typically quite small, about $6,000 in most programs. In the recent past, NSF has given them quite freely. If you submitted a request that seemed reasonable, you got one. However, things have changed recently. They are not reserved to "unanticipated" involvement of undergraduate students in your research. If you plan to include undergraduate researchers, you should request REU funding in your original proposal (include it at participant support in the budget and describe it in the budget justification). If your grant was awarded before mid-2013, you can still use the old system, which in practice means applying for small REU increments annually. Each program has separate deadlines for this (March 1 in IOS, and December 1 in DEB). So, as always, you need to check with your program officer. The new REU solicitation can be found at

http://www.nsf.gov/publications/pub_summ.jsp?WT.z_pims _id=5517&ods_key=nsf13542.

RAHSS (Research Assistantships for High School Students)

This program provides support (often summer support) to involve a high school student in your ongoing research project, typically under $6,000 per student. In my experience, applications usually identify a particular student, although this is not strictly necessary. RAHSS supplements are available for any project funded in NSF's BIO directorate. The RAHSS program is described in what NSF calls a "Dear Colleague Letter" that you can find at

http://www.nsf.gov/pubs/2012/nsf12078/nsf12078.jsp.

The use of a Dear Colleague Letter to promote this program is interesting because it tells you that this is something that BIO is promoting and encouraging. Dear Colleague Letters are messages to the community that say, "We're really interested in applications of this type." Internally, they are also messages to program staff that they should consider these applications seriously.

RET (Research Experiences for Teachers)

The RET program is similar to the REU program, except, of course, that it supports the involvement of K–12 teachers in your research rather than undergraduates. In fact, the RET program is actually part of the REU program and is governed by the rules laid out in the REU solicitation (see link above). Support is generally around $15,000 per teacher participant, and you can request support for two teachers. Proposals typically identify the teachers involved and provide an NSF-style CV for each participant. The BIO version of the RET program is described in a Dear Colleague Letter at

http://www.nsf.gov/pubs/2012/nsf12075/nsf12075.jsp.

ROA (Research Opportunity Award)

These supplements are designed to support the involvement of a faculty member from a primarily undergraduate institution in your project. These awards can be significantly larger than the other supplements, typically around $25,000. This mechanism is underused, and it deserves to be used more frequently because the right investigator from a small college can make a much bigger impact on your project than an undergraduate or a high school student. These funds can be used to involve the RUI-investigator during the summer to partially support the investigator's sabbatical. A Dear Colleague Letter describes the ROA program at

http://www.nsf.gov/pubs/2007/nsf07041/nsf07041.jsp.

Proviso.

Note that BIO, and IOS in particular, is trying to discourage the use of these supplements for things that you could have reasonably anticipated when you wrote your proposal. In the case of the REU program, this is now codified in the REU solicitation as I explained above. So you must now request REU support in your original proposal. In the case of the other "R" supplements, however, you will not make an argument that you could not have reasonably anticipated the need for an ROA, RET, or RAHSS in your original proposal.

General supplements

While the "R" supplements represent some of NSF's specific goals, the supplement is a very broad mechanism that gives NSF and its program officers considerable flexibility to support science. NSF can use supplemental funding in a great many ways. For example, NSF can provide a supplement of up to 20 percent of your current award (this could be in the neighborhood of $60,000 to $80,000 on a normal $400,000 award) if the program officer agrees with you that this supplement is needed to "assure adequate completion of the scope of the proposed research." This could arise if you overspent for some reason on a preliminary part of your project, so that you find yourself low on cash as you home in on the big questions that you planned to answer. Another possibility is that a new technique or analysis has become available that will answer your original question more accurately or completely. As with so many things at NSF, if you feel that your situation merits a general supplement, you need to speak to your program officer. Your success will depend largely on the program officer's judgment, so you want to be sure that your project and your request make a favorable impression.

I offer two examples to further illustrate the flexibility of the supplement mechanism. First, program officers can recommend a supplement in the form of a "special creativity extension." These supplements add up to two years of support to a three-year project that has been especially meritorious. These are not just "no-cost extensions" but an extra two years of cash support. In theory, program officers initiate these in response to annual reports, but there is nothing to prevent a principal investigator from drawing the program officer's attention to his or her accomplishments. Secondly, Dear Colleague Letters often rely on the very general supplement mechanism to promote one of NSF's goals. For example, IOS currently has an active Dear Colleague Letter that encourages older investigators with active awards to obtain further training in advanced techniques such as genomics. Although IOS calls these "Mid-Career Awards," they are basically supplements.

Although the "R" supplements (REU, RET, RAHSS, ROA) are reasonably well known, only the REU program is widely used, and it is now shifting away from the supplement mechanism. The others are relatively underused and merit your attention. The ROA and "general" supplements are most likely to have a big impact on your project. You should consider them.

Final thoughts: your plan of action.

Writing a compelling four page preproposal will challenge every potential investigator. In preproposal writing, less space means you must expend more thought per word. To achieve preproposal Zen, you will actually spend relatively little time putting words on paper. Writing a good preproposal is 80% preparation. Choosing the right idea and targeting the right NSF program are fundamental to this preparation. You preparation must not stop here, however, your preproposal must follow an outline that anticipates the issues panelists and program officers need to know. It follows that the better you understand this audience, the better you can speak to them. Finally, you need a plan for after the invitation decision. This involves decoding feedback you receive, knowing how you should interact with NSF staff, and understanding alternative funding mechanisms at the foundation.

Good luck!

Feedback please. Somebody once said that authors should 'use outside readers.' Seems like a good idea. If you have thoughts about how I can make this book more useful, or insights about good and bad NSF strategy, I'm anxious to hear from you. Email me at **stephxy@gmail.com**.

Getting more. To order more copies of this book please visit **www.preproposalzen.com**, where you can also get information about personal or group consulting, read my pre-proposal blog and find links to other useful information.

Appendix. Example of a submitted preproposal

I submitted this preproposal in January 2013. NSF invited a full proposal, so it was at least partially successful. The full proposal is currently pending.

This example follows the advice given here. Although you will see that I wrote a slightly longer background section than recommended here, because this gave me the opportunity to explain the proposed procedures, and therefore reduce the size of the research design section. The pre-proposal panel (quite correctly) criticized the brevity of the broader impacts, but they were otherwise enthusiastic about this idea.

Project Summary: Stephens Preproposal submitted

Overview. Communication is a basic attribute of human and non-human social interactions, and effective communication clearly requires honest signaling. Honesty, however, is conceptually problematic because it can be unstable when a conflict of interest exists between signalers and receivers. While investigators have proposed several theoretical solutions to this problem, the claim that the costs of signals enforce honesty, sometimes called the handicap principle, is probably the single most influential model of honest signaling. Yet the handicap hypothesis remains controversial in many ways. Some critics complain about weaknesses in the data, while others simply claim that other ideas (such as indices) are more parsimonious. Complaints about the data arise because most of the supporting evidence simply shows that signals are costly, which is necessary but far from sufficient to stabilize honesty via a handicap. What is clearly lacking is unambiguous experimental data in which investigator-controlled costs demonstrably enforce honesty. This project will conduct experiments of this type, pursuing three experiment aims: (i) characterize the effects of state dependent costs on signal honesty; (ii) explore the interaction between signaling costs and signaling benefits in the control of honesty; (iii) determine the role of environmental uncertainty in stabilizing or de-stabilizing honest signaling.

Intellectual merit. The proposed project has intellectual merit because experimental manipulation is the gold standard of empirical inference, yet it is very difficult to manipulate both equilibrium and non-equilibrium signaling costs in natural signaling systems. A rigorous experimental determination of

94

the role of signal costs and related variables has the potential to re-focus research programs that are exploring the many costs of naturally occurring signals. Moreover, empirical documentation of which costs are important and under what conditions is a key step in placing the handicap principle in its proper place with other explanations of honest signaling such as mutualism, indices and punishment.

Broader impacts. This project will promote the training of both graduate and undergraduate students by engaging them in meaningful research. Undergraduate students from diverse academic backgrounds will work both individually and as a team to conduct research in this project. In addition, the project will recruit high school and community college students from under-represented groups for a series of summer research programs. To achieve this, the investigator will develop connections with minority-serving high schools in the Twin Cities, and with Tribal Community colleges in the upper Midwest. In addition, the project will engage in public outreach via presentations at regional museums and nature centers.

Project Description, Stephens Submitted Preproposal

Note: The project description was five pages long as submitted (one page for personnel and four pages for text, as required). Reformatting this for publication may cause it to run longer.

Personnel

David W. Stephens

PI, Professor of Ecology, Evolution & Behavior, University of Minnesota.

Stephens will serve as principal investigator, establishing the basic procedures and supervising the day-to-day conduct of the proposed research.

1. Specific Aims.

1. Specific Aims. Signal reliability raises fundamental questions about animal communication. When a conflict of interest exists between a signaler and a receiver, honest signals should be unstable, because the signaler benefits from signaling dishonestly and the receiver benefits from ignoring the unreliable signal. While investigators have proposed many solutions to this problem, the claim that signaling costs somehow enforce signal reliability is one of the most widely invoked explanations of reliable signaling. It follows that an inferentially strong test of this claim would show that honesty persists in the presence of appropriate costs, but disappears when these costs are absent. Logical as this may be, the difficulties associated with such a program are significant. First, because the stability of reliable signaling is a game theoretical problem, it is not sufficient to measure or manipulate the costs that occur at equilibrium; the costs that occur in rare-but-important non-equilibrium situations are critical. Second, unambiguously characterizing a signal as honest or dishonest implies that the investigator has considerable knowledge about the information encoded in a given signal. In practice, our knowledge of the uncertain states that animals "signal about" is often no more than an educated guess. Clearly, we need to resolve these outstanding issues in order to move beyond controversies about the importance of costs in signaling, and hence to place the various mechanisms that may generate honesty in proper perspective.

The long-term goal of this research program is to develop a conceptually coherent, empirically well-supported approach to the role of costs in signaling that places signaling within the more general context of animal information processing. The objective of this proposal is to experimentally evaluate the effects of different types of "signal cost" on signal honesty using a novel laboratory system. The central hypothesis of this research holds that difficult-to-observe costs that arise when the interests of signalers and receivers are in conflict promote reliable signaling, while many easily observed costs have little or no effect. This hypothesis flows from an experimentally tractable variant of Maynard Smith's Sir Philip Sidney game (Maynard Smith, 1991). Our experimental approach will create a signal-receiver game in the laboratory using blue jays as experimental subjects. Within this highly controlled structure we can ask whether learning and other general mechanisms of behavior are consistent with the claims of signaling models. The rationale of this approach is two-fold. First, although models of signaling often emphasize genetically determined signaling traits, such as morphological ornaments, experience clearly plays a key role in both the production of natural signals (e.g. bird song) and in the responses of receivers to signals. Second, these techniques give us the control necessary to manipulate all the conceptually important variables in game theoretical models of signaling, such as the equilibrium and non-equilibrium costs of signaling, and the economic consequences of receiver actions. Moreover, because we control the properties of the game "signaled about" we have a clear-cut & operational definition of honesty.

We will accomplish the objective of this proposal by pursuing three aims:

Aim#1. Experimentally characterize the effects of state-dependent costs on signal honesty. We expect that the non-equilibrium costs of signaling are the primary determinants of signal honesty. This has intellectual merit because these are the most difficult costs to measure or manipulate in natural signaling systems.

Aim#2. Experimentally explore the interaction between costs and benefits in the behavioral control of honesty. We expect that the degree of economic conflict between signaler and receiver determines the importance of costs. This has intellectual merit because it speaks to the relationship between "cost-enforced honesty" and simpler mutualistic honesty.

Aim#3. Experimentally determine the role of uncertainty in the stability of honest signaling. Finally, we expect to support for the intuitive notion that the underlying frequency of states (uncertainty) can make it easier or harder to lie. This has intellectual merit because the central role of uncertainty in signaling systems is not widely appreciated.

2. Background & Significance.

The claim that costs promote honesty has a long history (see Bradbury & Vehrencamp, 2011; Maynard Smith & Harper, 2003; & Searcy & Nowicki, 2005 for reviews). Zahavi's (1975 1977) verbal formulation was met with dismissal and skepticism that lasted nearly 20 years until Grafen (1990) and Maynard Smith (1991) offered mathematically rigorous models of Zahavi's claim, and these models were quickly followed by others (e.g. Getty, 1998; Hurd, 1995; Johnstone & Grafen, 1993; Vega-Redondo & Hasson, 1993). In the wake of these algebraic results the conceptual viability of the handicap idea is now widely accepted; yet considerable controversy about the empirical validity, importance and correct interpretation of the handicap idea remains. These criticisms take two forms. First, critics claim that many tests of the model are flawed because they measure the wrong costs (Grose, 2011; Lachmann et al, 2001; Számadó, 2011, 2012). Second, given the lack of unequivocal data, the critics argue that simpler "honesty mechanisms" (such as conventions, mutualism, punishment or indices) are more parsimonious (e.g. Guilford & Dawkins, 1991; Hurd, 1995; Vehrencamp, 2000). *The research proposed here has intellectual merit because it will address both of these criticisms by conducting a conceptually sophisticated, inferentially strong experimental evaluation of the role of costs in signal honesty.* Notwithstanding the unresolved questions about how to properly measure costs and the relative importance of costs in the maintenance of honesty, the "costs of signaling" literature continues to grow in numbers and sophistication. A direct experimental analysis, as proposed here, has enormous potential to re-direct the energies of this field and focus its attention on the most promising types of costs, the most important alternative accounts, and ultimately to focus our attention on those physiological, neural, genetic and hormonal mechanisms that are more significant in generating reliable communication.

Table 1A. Receiver's Payoffs

Receiver Action	State	
	Good	Bad
Accept	1	0
Reject	0	1

Table 1B. Signaler's Payoffs

Receiver Action	State	
	Good	Bad
Accept	1	a
Reject	0	b

Table 1C. State Dependent Signaling Costs

Signal Emitted	State	
	Good	Bad
Signal	Honest c_1	Dishonest c_2
No Signal	Dishonest c_3	Honest c_4

3. Model & Hypotheses.

We imagine that some aspect of the environment exists in two possible *states* that we call "good" and "bad." We assume that the state is good with probability p. For any given play of this game, the signaler knows the state but the receiver does not. The receiver must choose between two alternative *actions* that we call "accept" and "reject." The receiver does best by matching its action to the state of the environment; specifically, "accepting" is best when the state is good and "rejecting" is best when the state is bad (Table 1A). Table 1B shows how the receiver's actions interact with the state to determine the signaler's benefits: When the state is good, "accept" is in the best interest of both players. There can be a conflict of interest, however, if the state is "bad." The receiver always benefits from rejecting a "bad" state, but the signaler's best interests are determined by the parameters a and b. If $b=1$ & $a=0$ we have a mutualism where the signaler's and receiver's interests are aligned; if, however, $b=0$ & $a=1$ we have a conflict of interest where the signaler benefits if the receiver "accepts," but rejecting is best for the receiver. Finally, we imagine that the signaler can choose between two actions that we call "Signal"--meaning indicate the good state-- and "No Signal" (which indicates the bad state). Note that while the terms used here are convenient, formally these are arbitrary labels for three general types of entities: states (here: good vs bad); receiver actions, (here: accept vs. reject); and possible signals, (here:

signal vs. not). Table 1C shows the costs paid by the signaler to emit these two signal types in the two environmental states. Notice that two of these state/signal combinations are honest and two are dishonest, so that c_1 and c_4 represent two different costs of honest signaling, while c_2 and c_3 represent the costs for two types of dishonest signaling. Table 2 shows a simple game matrix derived from these assumptions. The handicap principle emerges here: honest signaling can only be stable when the difference between the cost of *dishonestly* signaling "Good" (c_2) and the cost of *honestly* signaling "Bad" (c_4) exceeds the difference b-a, which measures the conflict of interest between signaler and receiver. Notice that the two costs c_1 and c_3 are irrelevant in this analysis, although c_1 is readily and commonly measured in natural signaling situations.

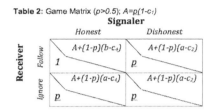

Table 2: Game Matrix ($p>0.5$); $A=p(1-c_1)$

		Signaler	
		Honest	*Dishonest*
Receiver / *Follow*	1	$A+(1-p)(b-c_4)$	$A+(1-p)(a-c_2)$
			p
Ignore	p	$A+(1-p)(a-c_4)$	$A+(1-p)(a-c_2)$
			p

4. Experimental Approach & Preliminary

Data. Our approach combines two techniques developed in our laboratory: experimental game theory (Clements & Stephens 1995; Stephens et al 2002) and studies of signaling economics (McLinn & Stephens 2006, 2010). Following our preliminary studies, we will house pairs of blue jays (*Cyanocitta cristata*) in side-by-side chambers, one for the signaler and one for the receiver (Fig. 1). The following paragraphs explain how we use this apparatus to implement the game outlined above while systematically manipulating each of the relevant variables.

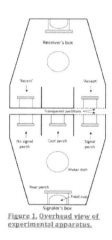

Figure 1. Overhead view of experimental apparatus.

Uncertainty & States: The signaler and receiver play the signaling game in a sequence of repeated trials. At the beginning of each trial, the computer determines the "state of the environment" ("good" or "bad" as described above) according to probability p (which measures uncertainty). The state in a given trial is unknown to the receiver but cue lights privately indicate the state to the signaler. In each trial, the receiver must choose between one of two stations at the front of its chamber, and the signaler may choose to signal about the state of the trial.

Payoff Consequences. We can easily program the apparatus so that the receiver's choice determines the amount of food delivered to both players as in Tables 1A & C. For example, we can create a conflict of interest ($a=1$, $b=0$) by delivering no food to the signaler if the receiver chooses "reject" in the bad state ($b=0$), but delivering some food when the receiver choose "accept" in the good state ($a=1$).

Costs: The signaler is faced with three stations; two that represent signaling states ("signal" and "no signal"), and a central "cost" station. Using this arrangement we can manipulate the Table 1C's four costs of signaling by requiring that the signaler complete a "hop requirement" (hopping back and forth between the cost station and the rear station a fixed number of times) that depends, in general, on the signal emitted by the signaler and state of the environment selected randomly by the apparatus.

Preliminary Data. In the pilot data (Fig. 2), we tested a factorial combination of two payoff condi-tions ("Interests Aligned," and "Interests in Conflict") and four levels of "signaling cost" ($c_1=c_2=0$ hops, 1 hop, 4 hops or 7 hops; $c_3=c_4=0$ in all conditions). Although this cost structure represents a common assumption for "handicap models," it confounds the effects of c_1 and c_2, which are quite different in theory. Figure 2 shows a clear-cut result. In mutualism the signaler signals honestly regardless of costs as we expect. When a conflict of interest exists, however, honesty depends on the costs of signaling as the handicap principle

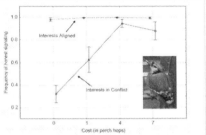

Figure 2. The effect of cost on honest signaling (Mean ± SE) for two payoff conditions (interests aligned---$a=1$, $b=3$, table 1B---and interests in conflict---$a=3$, $b=1$) and four levels of signal cost ($=c_1=c_2$; $c_3=c_4=0$, table 1C; Polnaszek & Stephens in prep). The Y-axis shows the relative frequency with which the signaler honestly chose the "no signal" perch when "reject" was the receiver's best option. [$F_{(3,18)}=14.0$, $p=0.00006$, repeated measures ANOVA]. METHOD: The data are from 14 blue jays randomly assigned to seven sig-naler/receiver pairs. Each pair experienced each "payoff condition by cost" treatment in a different randomly determined order for 900 free trials; the data here are from the final 300 of these trials. 10% of trials were "forced" trials designed to ensure that the subjects had ongoing experience with all options. By design, signaler-receiver pairs began each treatment with the signaler honestly signaling and the receiver following, which we engineered by exposing pairs to a mutualism "pre-treatment." So, these data represent a test of stability as the logic of game theory requires.

predicts. The proposed research will use these procedures to pursue three aims as outlined below.

Aim#1. Experimentally characterize the effects of state-dependent costs on signal honestly. The central hypothesis is that the relationship between c_2 and c_4 enforces signal honesty while c_1 and c_3 have no effect. The experiments proposed for this aim are straightforward extensions of our preliminary results involving factorial manipulations of the four signaling costs.

Aim#2. Experimentally explore the interaction between costs and payoffs in the behavioral control of honesty. The central hypothesis is that the costs required to enforce honesty depend on the payoff structure of the signaling game (i.e., the conflict of interest variables *a* & *b*). Guided by the results of Aim#1, the empirical approach for this aim would be to factorially manipulate costs and the payoff matrices of both the signaler and the receiver.

Aim#3. Experimentally determine the role of uncertainty in the stability of honest signaling. The central hypothesis for this aim is that underlying uncertainty (or base rate) about "state" can either promote or destabilize signaling. Intuitively, it is easier to lie about common states. This is an under-appreciated issue in the study of signaling. The experiments proposed for this aim will manipulate the variable *p* across a range of payoff and cost conditions. We expect that the power of a given cost condition to stabilize honesty depends on the degree of uncertainty.

Summary. Our results will represent an important step in understanding the forces that make communication reliable. Specifically, our results have the potential to re-shape and re-direct the ongoing research enterprise that focuses on the mechanisms that generate signal costs whether these putative costs are social, immunological or developmental.

5. Broader impacts. The Stephens lab has a long history of involving undergraduate students, high school students and high school teachers. In annual departmental evaluations our research group consistently ranks #1 or #2 in the number of undergraduate advisees. For the purposes of this project, we will recruit a group of 5-6 undergraduates from a diverse set of academic and social backgrounds. These students will work together to assist in the completion of the aims of this project. They will attend weekly lab meetings to coordinate work and to read papers from the primary literature. Each student will be supervised by a "lab mentor," which will be the PI, a postdoc or a senior graduate student. We will conduct a similar summer program for advanced high school and community college students. We will recruit students from the Twin Cities' large Somali and Hmong populations through contacts with science teachers at high schools with large populations of these students. Moreover, we will contact science departments at each of the 13 Tribal Community colleges in the upper Midwest (ND, SD, MN, & WI).

Made in the USA
Charleston, SC
18 July 2014